The
Young Oxford Book
of
Ghost Stories

Other Anthologies edited by
Dennis Pepper

The Oxford Book of Animal Stories

An Oxford Book of Christmas Stories

The Oxford Christmas Storybook

The Oxford Book of Scarytales

The *Happy Birthday* Book
(*with David Jackson*)

A Book of Tall Stories

The
Young Oxford Book
of
Ghost Stories

DENNIS PEPPER

OXFORD UNIVERSITY PRESS
Oxford • New York • Toronto

Oxford University Press, Walton Street, Oxford OX2 6DP

Oxford • New York • Toronto
Delhi • Bombay • Calcutta • Madras • Karachi
Kuala Lumpur • Singapore • Hong Kong • Tokyo
Nairobi • Dar es Salaam • Cape Town
Melbourne • Auckland • Madrid
and associated companies in
Berlin • Ibadan

Oxford is a trade mark of Oxford University Press

This selection and arrangement © Dennis Pepper 1994
First published 1994
Dennis Pepper has asserted his moral right to be identified as the author of this
work

A CIP catalogue record for this book is available
from the British Library

ISBN 0 19 278126 X

Printed and bound in Great Britain by
Butler & Tanner Ltd, Frome and London
Typeset in Stempel Garamond
by Pentacor PLC, High Wycombe, Bucks

Contents

Introduction

When I started to collect stories for this book I thought I knew what a ghost was: the spirit of a dead person that returns to haunt the living. That's a ghost all right, but now I think there's more to ghosts than first meets the eye.

Even this sort of ghost can assume various forms. It might look so like the living person that you can't tell it's a ghost, or it might be, well, ghostly—semi-transparent. It might even be totally invisible, apparent only through its actions or the feeling of intense coldness that accompanies it. It might be visible to some people but not to others, or it might change the form of its appearance during any particular manifestation. However, the more I read the more I became aware of ghosts of a different kind. There are, for instance, stories about ghosts that don't know they're ghosts, and about the ghosts of people who are still very much alive. There are, too, ghosts that exist (as perhaps all ghosts do) only in the minds of those who believe in them.

I was finding more ghosts than I had bargained for, but it didn't stop here. People claim to have had experience of spectral dogs and cats, ghostly horses and owls and monkeys, and — no doubt—many other animals. These appear in stories too. As well as ghosts that are the spirits of dead *people* there are, then, other creatures whose apparitions return after death.

It doesn't stop here, either. There are ghostly objects: phantom coaches (and buses) pick up passengers, Second World War aeroplanes land on long-disused airfields, ghostly steam trains run along abandoned tracks. There are ghost ships and cars. And people have stayed at inns that have been in ruins for longer than anyone can remember, visited villages that no longer exist, attended services in churches now derelict. They have heard, and sometimes seen, battles that were fought centuries ago. *Things*, too, can be ghostly.

What is the point of all this ghostly activity? In ghost stories, as in recorded experiences, apparitions of the dead haunt the living for various reasons. They may appear because they cannot rest until they have revealed a crime and seen that the person who committed it is punished; or they may be unable to leave the scene of a particularly horrible crime—a murder, say—in which they have been involved, either as the criminal or as the victim. They may have been buried in unhallowed ground and wish to be reburied where they can rest in peace. They may return to keep a promise or fulfil a curse, to guard or protect or help someone, to reveal hidden treasure, to warn of forthcoming events. Others may return to take part in some long-forgotten ritual or to seek vengeance themselves rather than urge the living to seek it on their behalf. There are, too, ghosts that are simply malevolent: very nasty pieces of work out to make things really unpleasant for any living person who is in the wrong place at the wrong time.

The stories in this collection are about ghosts like these, but even so it is sometimes difficult to know whether you are dealing with a ghost or some other kind of supernatural entity (of which there are many). When the corpse of a long-dead person refuses to stay buried, I don't know whether it is the ghost of the once-living person which is animating it, or not. What do you think? And since ghosts are, mostly, 'out-of-time' —beings from the past that haunt our present—can some of the characters in other 'time-slip' stories be said to be ghosts? What do you think? You will find both kinds in this book.

Many of the finest modern ghost stories are by writers such as Vivien Alcock, Jane Gardam, Adèle Geras, John Gordon, Jan Mark, Philippa Pearce, Alison Prince, Robert Westall—the list is longer than this—all of whom write especially for young readers. But I thought it was important to place alongside their work stories by authors who write mainly for adults, both those from an older tradition of ghost-story writing—E. F. Benson, Marjorie Bowen, Bernard Capes, M. R. James, E. Nesbit, for instance—as well as more modern authors—Ruskin Bond, Stephen Dunstone, Gerald Kersh, Elizabeth Walter, and Mary Williams. A book for young readers is very limited if it is confined to stories especially written for them.

Now it is time to switch off the light, get out your torch and be prepar—Sh! What's that?

Dennis Pepper
April 1994

The Rivals

VIVIEN ALCOCK

John Pearce was a clever boy. Every Speech Day would find him, blinking modestly behind his thick spectacles, trotting up to the platform to receive prize after prize into his thin, eager hands. Everyone said he would go far—though probably not on his feet, which were flat as Dover soles, and inclined to smell fishy in hot weather. His English teacher, sounding the only sharp note in a chorus of praise, said he had no imagination, but this was not quite fair. John believed in many things he had not seen: atoms and molecules, microbes and magnetic fields. He did not, however, believe in ghosts.

So when the milkman told them, the morning after he and his parents had moved into their new home, that the house next door was haunted, he laughed and said,

'What rot! I don't believe in ghosts.'

'John is so sensible,' his mother said. She stepped outside and peered at the houses on either side. 'Which one?' she asked, with an enjoyable shiver.

It was, of course, the one on the left, the sinister side: a dark house, the colour of dried cat's meat, its sly, gothic windows masked with heavy Nottingham lace. No clear light of day or reason could penetrate those dingy cotton flowers and flourishes into the hidden rooms behind. Put in some picture windows, and white nylon net, John thought, and there'd be no more talk of hauntings.

The house had changed hands six times in the last three years, the milkman told them, handing over two bottles, white as ghosts. No one could stand it any longer. All that screaming and wailing, footsteps and icy draughts of air. Cost a fortune just trying to keep it warm! The present owners had been there only five weeks. 'Got it dirt cheap,' he said. 'Thought they were getting a bargain, poor devils. I give them six months.'

After breakfast, Mrs Pearce shooed John out into the garden, saying she could manage better on her own. No, he was not to go up to his room and stick his nose into a book. He worked too hard at his studies. The sunlight and fresh air would do him good. He was looking pale.

John wandered down the neat concrete path, examined the aphids on the roses with a critical eye, murmured the Latin names of all the plants he recognized, and made a note to look up those he did not. Then, with nothing left to do, he climbed onto a tree stump and looked over the high wall into the garden next door.

On a neglected lawn, ankle-deep in daisies and dandelions, a young girl was standing, bouncing a ball against the thick trunk of a chestnut tree. A pretty girl, with a face like a flower and long dark curling hair.

His heart sank a little: he could have wished she were plain. It was not that he disliked pretty girls, far from it, but he knew from experience that they did not like him. His learning did not impress them. His clever remarks, his carefully prepared jokes, fell flat. They giggled behind their hands, called him Four Eyes, and yawned in his face when he tried to share his knowledge. All the time he was speaking to them, he could see their eyes skittering across the classroom to some good-looking thick-head on the other side. Pretty girls, he thought glumly, were always stupid.

Yet he was so lonely. He would have liked a friend. Perhaps she had brothers? He thought hopefully of a quiet, studious boy, like himself. They could go round the Science Museum together, not just rushing around pressing buttons to make things light up, but slowly and seriously. They could collect pondlife in jam jars, and study it under his microscope. They could have picnics in his room, discussing the theories of Pythagoras by candlelight. If only she had such a brother . . .

'Hallo,' he called.

The girl started and dropped her ball, which vanished in the long grass. Then she smiled and came towards him. 'Hallo,' she said.

'I'm John Pearce. We've just moved in.'

'Yes. I know.'

'What's your name?'

'Lucy Wilkins.'

There was a little silence. She seemed shy. He studied her. Her eyes were remarkable: a pale, sparkling blue, cool as water, and fringed with long black lashes. Her voice was clear and sweet, and rather posh. Perhaps she went to a boarding school and was lonely in the holidays. Perhaps she too would be glad of a friend. It was a pity she was pretty, and, QED, bound to be stupid.

He asked her if she had any brothers or sisters, and was disappointed when she shook her head.

'I'm an only child too,' he said. And for a moment, because she was so pretty, he hoped (foolishly, he knew) that she would say, 'Let's be friends. I've always wanted a brother. It's lonely on your own.' But she did not, of course. She simply smiled and said nothing. Already she was looking bored, and her eyes were sliding away from him, glancing back at the dark house as if even its gloomy privacy would be better than his company. Any moment she would make an excuse—she had to help with the dishes or the dusting, or wash her hair. He wished he could think of something to say to keep her there . . .

'Have you seen any good ghosts recently?' he asked humorously.

'Oh, you've heard already!' she said crossly. 'Who told you that—'

'That your house was supposed to be haunted? The milkman. Icy draughts, strange wailings, footsteps in the night; all the usual old rubbish . . . You don't mean you believe in it, do you?' he asked.

She stared back at the dark house and shivered. 'Yes.'

Poor silly girl, he thought. He explained to her kindly that there were no such things as ghosts. Icy draughts in an old house were only to be expected. 'You should buy some plastic filler and seal up the cracks.'

She looked sulky and mumbled, 'You don't know what it's like.'

As for the footsteps . . . He went on to explain about the expansion and contraction of old floorboards, affected by temperature and humidity.

She sniffed, and stuck out her lower lip.

And the wailing, he said, that would be wind in the chimneys.

It was funny. You would think people would be glad to have their fears and worries explained away in a rational manner, but they never were. He was not surprised when, instead of looking grateful, she merely scowled.

'It is haunted! It is! I know it is!' she said stubbornly.

He laughed. 'Have you ever seen a ghost? Actually *seen* one?'

'*Yes!*'

The little liar! 'What did it look like?'

She hesitated.

Caught her there. She's got no more imagination than I have, John thought with satisfaction.

'Oh, horrible, horrible,' she muttered at last, obviously unable to think of anything better. 'Wicked! And it's me it's after. I know it is! It wants to drive me out . . . Oh, it's easy for you to laugh! You're safe next door.'

'I suppose it walks when the moon is full?'

She nodded. 'In that room,' she said, pointing to a top window overlooking the garden. 'At midnight, that's when it comes. Searching, looking for *me!*'

'Why don't you just lock the door and shut it in?'

'There's no key,' she said, and her voice trembled, 'for a door like that.'

Poor silly, pretty little fool! She really was frightened, he thought, and his heart filled with a warm, protective love that he had only felt once before, when he had seen a little white mouse in his father's laboratory. Wanting to comfort her, he offered to stay all night in the haunted room.

'I'm not afraid,' he said, and she looked at him with huge eyes, as if unable to believe anyone could be so brave. Or so foolish.

'The moon is full tonight,' she whispered.

They decided not to tell their parents. Parents, John informed her, could never be relied on not to produce objections to the most innocent and harmless schemes.

At ten to midnight, John knocked softly on the side door of the haunted house. It opened immediately. She must have been waiting behind it. Her face was as pale as a lily in the shadows.

'Come,' she whispered, and led the way upstairs. Softly though he trod, the stairboards creaked and groaned beneath

his feet. If anyone hears, John thought, they'll take us for ghosts. But he didn't want to be caught. It might be difficult to explain . . .

'Have your parents gone to bed yet?' he whispered.

'Yes. A long time ago.'

They went up two flights of stairs, along a narrow passage, and then Lucy opened a door. It creaked. The hinges need oiling, he thought.

Now they were in a large room. Bright moonlight struggled through the thick lace curtains, patterning the floor with wriggling little worms of light. John tiptoed across to the window and pushed them back.

'That's better,' he said, and looked round. There was no furniture, and the floor was carpeted only with dust. The wallpaper was dark. Opposite the window, a large mottled mirror, in a heavy frame, gleamed dully in the moonlight. On the left, there was a pale marble fireplace, an empty grate, and two cupboards in the alcoves on either side. He opened each door in turn and shone his torch inside. Empty. Dirty. Dusty.

'No wonder your ghost only visits once a month,' he whispered, grinning, 'if this is the room you give it.'

Lucy did not answer. She was sitting in a corner, with her arms wrapped around her. He could see the whites of her eyes as she kept glancing nervously from the door to the window and back again.

In the garden, an owl hooted. It was very cold. Strangely cold for a summer night.

'That's because it's a corner house,' John explained. 'It catches the wind both ways. Draughty.' And he told her about air currents and wind velocity. He did not know if she was listening. Her eyes still moved from the door to the window, and back again.

A clock began to strike twelve. The curtains flared wildly at the window. The house creaked and shuddered. There was a thin wailing from the garden below.

'The wind's coming up,' John said. 'They said on the radio the weather was going to change.'

Now there were screams, wild, despairing, unearthly.

'Cats,' John said, and began to tell Lucy about the mating habits and aggressive displays of cats . . .

'Look!' she whispered.

The door was opening, slowly, slowly. A figure appeared. Thin as a candle, it flickered into the moonlight. Its face was grey and gaunt, its white gown all spattered and splashed with blood.

'How d'you do,' John said, getting to his feet and blinking at it short-sightedly. 'Are you Lucy's mother?'

It drifted towards him, moaning and wringing its hands. Its eyes were burning like coals in its ashen face.

'Aren't you feeling well?' John asked, uneasily. He took off his glasses and wiped them on his handkerchief. But when he put them on again, there was no improvement. The lady (for it appeared to be a female) looked dreadful. 'Can we get you anything? An aspirin?' He looked towards Lucy for help, but she was cowering in her dark corner, and did not move.

'Oh, I am murdered, murdered!' the lady wailed. 'Murdered in my bed!'

A joke! A practical joke! They had planned it together to make a fool of him.

'Ha! Ha! Very funny,' he said, furious that they should have thought him so gullible. 'But I'm sorry. It's wasted on me. I don't believe in ghosts. And I'm afraid it's time I was getting back. I promised to help Mum in the morning.'

'Murdered!' the lady repeated, staring at him with hollow eyes. 'Murdered in my bed, the wicked devils!'

He stared back at her stolidly, refusing to be frightened, and a look of impatience came into her ravaged face. 'Murdered,' she repeated slowly, as if to a backward child. 'October the second, it was, in the year of disgrace, 1872.'

'That's a long time ago. I should forget it if I were you,' John said stoutly, and was surprised how high his voice sounded, almost like a scream. It seemed a long way to take a practical joke. It occurred to him that perhaps the lady was mad.

'Don't you believe me?' the apparition asked, her icy breath chilling his cheek.

'No,' he said. His pulse was racing now. He was burning and shivering. It wasn't fear, he told himself. He must be sickening for something.

'Look at me!' She came nearer, and John backed away until he was against the wall.

He did not want to look at her. His glasses must be misting up in the freezing air, and that was why her face seemed to be melting, dripping from her bones like candle wax.

'Oh, please! You must believe in me, you must!' the lady moaned. 'Even the gods die for want of faith. I need your fear. I can only exist in your mind. Oh, please believe in me, or I am lost.'

'I'm sorry,' John said stubbornly, his teeth chattering. 'I don't believe in ghosts. I don't! I won't!'

The figure seemed to dwindle, fading away like smoke.

'Please,' it wailed faintly. 'Oh, please believe in me.'

John was shaking all over now, but he managed to say, 'No.'

'Oh, I am murdered, murdered, murdered,' sighed the ghost, its voice failing. Then it was gone.

The boy, his back to the wall, slid slowly down till he was sitting on the floor. His face was white, fixed, terrified.

'She's gone! She's gone! You did it!' the girl cried, smiling and clapping her hands together.

'It—it was a trick!' he babbled. 'I know it was! You had a hidden projector! A video tape! It's just a trick! You had something up your sleeve!'

The girl was dancing on the floor. Her feet made no sound on the bare boards. They left no prints in the dust. Now she danced in front of the mirror, and there was no reflection. Her eyes were shining, literally shining, like twin lamps. As John watched her in terror, she cried gloatingly, 'It's mine! All mine now! The whole house! She's gone, that horrible creature. Always scolding. Always criticizing. Saying I shouldn't walk in the sun. Saying I didn't know how to haunt properly, just because she's been dead longer than I have. Why should I care for her silly rules? I'll show her! Oh, I'll be *ghastly*!'

She came dancing towards him, and she was all moonlight.

'Thank you, thank you, thank you,' she whispered. 'I love you.'

And she vanished.

As he sat staring at the empty room, he felt an icy touch on his lips, as soft and wet as a snowflake.

He never saw her again. The two houses were sold within the year, and a block of flats built where they had stood. John and his parents moved to the other side of town. He grew up, won more prizes, and became rich and famous, and happy enough. But he never married. Sometimes, on a summer night, he would stand by his window, and sniff the sweet smell of night-scented stocks, and see the pale roses glimmering in the moonlight. Then he would smile and say,

'A pretty girl once loved me.'

Spook House

MARC ALEXANDER

I'm going to the Spook House after school,' Christopher Etheridge told the class just before Miss Mason arrived for the next lesson. 'Who's got the bottle to come with me?'

There was a buzz of talk. All that morning the school had been whispering about the old house—believed to be haunted—which stood in a large overgrown garden in Richardson Avenue.

'Here comes Minnie,' cried the lookout at the door. When Miss Mason entered she was delighted to find that everyone had their eyes intent on their geography textbooks. During earlier lessons she had sensed an air of excitement among the pupils which happens in schools from time to time, an excitement which puzzles the teaching staff and remains a mystery to adults.

While the lesson droned on through the heat of the summer afternoon, Christopher congratulated himself on being the cause of today's excitement. It was he who had brought the story of the Spook House to school, and it had spread so fast that at morning break someone actually asked *him* in the locker room if he had heard about the ghostly goings on in 'that old Hillsdale place'. When the lesson was over, and everyone was banging their seats as they got ready to go home, several members of the class crowded round him.

'Chris, are you really going into that house?' asked Dawn Sheehan breathlessly. 'I wouldn't dare!'

'What's supposed to haunt it?' asked her friend, Mavis Tucker.

'A murder victim,' said Christopher in a low voice. 'A lady was murdered there, and her ghost comes out at night all white and gleaming—and thirsting for revenge. And they say that in the kitchen, where she was stabbed with a carving knife, there's a pool of blood that won't dry.'

Some of the children shivered delightedly at the thought. It was as good as a horror film on the telly.

'And what else?' asked Mavis.

'You want more—isn't that enough?' Christopher asked. 'Well, they do say that the murderer hid her body down in the cellar in a sort of grave, and people only found out what happened when a bony hand came up out of the ground.'

'What a load of old rubbish,' came a cheerful voice. Christopher turned to see a tall boy with glasses. His name was Jeff Dixon and he had never been one of Christopher's friends. 'Where did you get all this spook stuff from?'

'If you must know,' said Christopher with dignity, 'I heard about it from my brother who is a reporter on the *Herald*.'

'Then why isn't it written up in the *Herald*?' demanded Jeff. 'I'll tell you why—because there aren't such things as ghosts any more than there's Santa Claus.'

'My brother told me that the editor said the story was too frightening to print. It would be unfair to people living close by, they'd be so scared they'd want to leave their homes.'

'Strewth!' said someone in an awed voice.

'Nah, it's rubbish!'

'OK, OK. If you're so sure, you wouldn't be scared to go into the house. You'd come with us this afternoon. Right?'

'Right! Who else is coming?'

'We'll watch you go in,' said Dawn.

Christopher led the way to the school gate, a small group of children following. On the way to Richardson Avenue Jeff talked to the others about the Spurs v. Chelsea midweek match the night before. Christopher remained silent and prayed that the house would at least look scary so they would not laugh at him.

Richardson Avenue was a long quiet road which gently sloped up a small hill. There was not a single soul in sight and Christopher was quick to point out that the silence was 'uncanny'. Jeff snorted.

At last the children halted where a cluster of estate agents' FOR SALE signs swayed above an overgrown hedge and a corroded nameplate on the front gate told them they had reached HILLSDALE.

'See,' cried Christopher triumphantly. 'Those signs have been there for months—there's moss on some of them—and that proves no one wants to buy a haunted house.'

'In we go,' Jeff shouted. 'You'll see it's just an ordinary empty house.'

'Do you think you ought to?' Dawn asked. 'Wouldn't it be breaking and entering or something . . .'

'Nobody will see us with all these trees and bushes growing in the garden,' said Christopher. The gate let out an awe-inspiring screech as he opened it which made Mavis give a little scream. 'Hush,' he said. 'We don't want to disturb anything!'

To his relief there was something eerie about the silent house and its wilderness of a garden. Brambles clutched at the children's legs as they followed a path leading to the back, and rose bushes, which should have been pruned ages ago, scratched their arms. Their voices dropped to whispers as though they were afraid to break the hot silence which hung over the house. Everyone jumped when Jeff said in his normal voice, 'Show us how to get in, Chris.'

'Just a minute,' he answered, trying the back door which he found to be locked.

'It looks like we can't get in,' said a boy thankfully. Even on the outside the house was very gloomy. Its windows were dirty, and the drawn curtains gave them a blind look.

Mavis looked over the knee-high grass that was once a lawn. 'Must be a long time since anyone lived here,' she murmured. 'When did the murder take place?' At the word 'murder' the children felt their skin creep.

'Here,' called Christopher. 'This little window isn't shut properly. Bring that old bin over and I'll climb through.'

'Aren't you afraid that a skeleton hand will grab you as you climb in?' teased Jeff, but even he kept his voice down. Christopher turned the bin upside down and climbed on to it, opened the window and wriggled through.

'Gosh, he's brave,' said Dawn. A moment later the back door opened. Christopher appeared grinning.

'I only needed to slide back the bolt. Who's coming in?'

Slowly the children walked through the doorway to find themselves in a large kitchen.

'Is this where the lady was killed?' someone asked.

Jeff snatched a coloured handkerchief from Dawn's blazer pocket and threw it into the cellar. 'If I come in tonight and go down into the cellar and bring back that handkerchief—which Dawn will recognize as hers—I'll win the bet. OK?'

'Fair enough,' agreed Christopher. 'Let's meet at nine o'clock.'

With relief the children filed back into the kitchen. Christopher was the last to leave. When he was alone he turned back the front door lock and flipped on the catch.

Careful not to be seen, the children sneaked out of the garden. They need not have worried, though; the avenue was still deserted.

'Be seeing you then—nine o'clock,' said Jeff. He went off with some of the other children, while Christopher stayed outside the gate with his particular mates.

'Listen,' he said. 'Old Jeff is too full of himself. I've got an idea . . . '

Nine o'clock had just tolled from the nearby church tower when Jeff entered the garden of Hillsdale and, thanks to it being a clear night, found his way round to the back of the house. The wind was cold and he wore his anorak. In his hand he held a small electric torch.

'Anyone there?' he called in a low voice.

Several figures appeared from the shadow of a weeping willow.

'Hello,' hissed Christopher. 'I'm glad it's you and not me that's going in there. If you want to we'll call it off—no one would blame you.'

'Trying to put the frighteners on me,' laughed Jeff. 'Check your watches, gentlemen. I shall be back in ten minutes with Dawn's handkerchief.'

'Won't you change your mind?' asked Christopher in a worried tone. 'It'll be on my conscience for the rest of my life if the ghost strangles you with her bony fingers.'

'It's no good, Chris,' said Jeff with another laugh. 'You just have your fiver ready.'

A moment later he entered the house through the kitchen door.

Inside the dark kitchen Jeff took a folded sheet from under his anorak and hung it over himself, adjusting it so he could see out of the eyeholes he had cut. It had been jolly good of Dawn to phone him with a warning that Chris planned to dress up as the ghost to scare him. She had thought it was very unfair, but Jeff decided to turn the tables on Chris when he crept into the house through the front door.

He switched on his torch and found his way to the cellar entrance. He decided he might as well collect the handkerchief and descended the steps carefully, the beam of his torch playing on piles of dusty junk. In one corner of the cellar someone had been digging a long time ago, probably to mend a water pipe. Jeff bet that it was that hole which had inspired the ridiculous story. He picked up Dawn's handkerchief and then switched off the torch to wait for Christopher in the dark.

As soon as Jeff had vanished into the house, Christopher's friends draped him in a torn white curtain which his mother had thrown out.

'This'll teach old Jeff,' someone said as he poured a bottle of red ink down the material.

'I still think it's mean,' protested Dawn.

'I think you're getting soppy over him,' teased Christopher. 'Isn't that right, Mavis?'

Mavis giggled.

'OK, here I go—the Phantom of Hillsdale,' whispered Christopher. His white form faded away as he hurried round to the front door where he silently let himself in.

'There really is a ghost in there!'

In the shadowy back garden the children's eyes widened with alarm as Christopher came blundering round the side of the house.

'I—I saw it,' he gasped. 'When I made up the story I never thought . . . '

'For goodness' sake, calm down and tell us what happened,' Dawn said.

'I—I got in all right,' he stammered. 'I haven't got a torch so I had a box of matches. I lit one and had gone quite a way before it burnt my fingers. I was just lighting the next when—' his voice trembled '—when I saw a sort of whitish shape. It was floating towards me. It—it reached out with its arms.'

'And what happened then?' asked Mavis.

'I threw off my curtain and bolted. Luckily the front door was open. I can still see it—it was ghastly.'

The others looked anxiously at the house, half expecting to see a spectre glide towards them.

Suddenly Christopher gave an exclamation of horror. 'Jeff is still in there,' he cried. 'We must get him out.'

'I'm not going in,' said a boy. 'Let's shout from here.'

'He'll not hear us if he's in the cellar,' said Christopher. He swallowed hard. 'It's down to me to warn him,' he muttered and took a step towards the house.

Dawn burst into a peal of laughter. 'Serve you right, Chris. What you saw was Jeff. I thought your plan was so mean I rang him up and warned him. He said he'd double-trick you by pretending to be a spook. He had a sheet under his anorak.'

Christopher gave a sigh of relief and slumped on to a lichened garden seat. 'Thank goodness,' he breathed. 'I won't even mind handing over a fiver. I tell you, I've never been so scared.' He glanced at his watch. 'He'll be out in five minutes and then what say we go down town and have a McDonald's?'

'That's the best idea you've had today.'

The children settled down to wait.

Jeff was getting bored in the cellar. He switched on the torch briefly and his watch told him that he had been lurking there for five minutes. It seemed more like half an hour. When would that silly Chris come?

Ah, there was something.

He looked up and saw a vague, glimmering figure slowly descend the steps.

Uttering a banshee wail, Jeff jumped into the centre of the floor where Christopher would be able to see him in the faint light from the doorway.

To Jeff's disappointment Christopher did not turn tail.

'All right, Chris,' he said. 'I reckon we're quits.'

But, with arms outstretched, the pale figure floated towards him . . .

The House
— *with the* —
Brick-Kiln

E. F. BENSON

The hamlet of Trevor Major lies very lonely and sequestered in a hollow below the north side of the South Downs that stretch westward from Lewes, and run parallel with the coast. It is a hamlet of some three or four dozen inconsiderable houses and cottages much girt about with trees, but the big Norman church and the manor house which stands a little outside the village are evidence of a more conspicuous past.

This latter, except for a tenancy of rather less than three weeks, now four years ago, has stood unoccupied since the summer of 1896, and though it could be taken at a rent almost comically small, it is highly improbable that either of its last tenants, even if times were very bad, would think of passing a night in it again.

For myself—I was one of the tenants—I would far prefer living in a workhouse to inhabiting those low-pitched oak-panelled rooms, and I would sooner look from my garret windows on to the squalor and grime of Whitechapel than from the diamond-shaped and leaded panes of the Manor of Trevor Major on to the boskage of its cool thickets, and the glimmering of its clear chalk streams where the quick trout glance among the waving water-weeds and over the chalk and gravel of its sliding rapids.

It was the news of these trout that led Jack Singleton and myself to take the house for the month between mid-May and mid-June, but as I have already mentioned a short three weeks was all the time we passed there, and we had more than a week of our tenancy yet unexpired when we left the place, though on the very last afternoon we enjoyed the finest dry-fly fishing that has ever fallen to my lot.

Singleton had originally seen the advertisement of the house in a Sussex paper, with the statement that there was good dry-fly fishing belonging to it, but it was with but faint hopes of the reality of the dry-fly fishing that we went down to look at the place, since we had before this so often inspected depopulated ditches which were offered to the unwary under high-sounding titles. Yet after a half-hour's stroll by the stream, we went straight back to the agent, and before nightfall had taken it for a month with option of renewal.

We arrived accordingly from town at about five o'clock on a cloudless afternoon in May, and through the mists of horror that now stand between me and the remembrance of what occurred later, I cannot forget the exquisite loveliness of the impression then conveyed.

The garden, it is true, appeared to have been for years untended; weeds half-choked the gravel paths, and the flower-beds were a congestion of mingled wild and cultivated vegetations. It was set in a wall of mellowed brick, in which snap-dragon and stone-crop had found an anchorage to their liking, and beyond that there stood sentinel a ring of ancient pines in which the breeze made music as of a distant sea.

Outside that the ground sloped slightly downwards in a bank covered with a jungle of wild-rose to the stream that ran round three sides of the garden, and then followed a meandering course through the two big fields which lay towards the village. Over all this we had fishing-rights; above, the same rights extended for another quarter of a mile to the arched bridge over which there crossed the road which led to the house.

In this field above the house on the fourth side, where the ground had been embanked to carry the road, stood a brick-kiln in a ruinous state. A shallow pit, long overgrown with tall grasses and wild field-flowers, showed where the clay had been dug.

The house itself was long and narrow; entering, you passed direct into a square panelled hall, on the left of which was the dining-room which communicated with the passage leading to the kitchen and offices.

On the right of the hall were two excellent sitting-rooms looking out, the one on to the gravel in front of the house, the other on to the garden. From the first of these you could see, through the gap in the pines by which the road approached the house, the brick-kiln of which I have already spoken.

An oak staircase went up from the hall, and round it ran a gallery on to which the three principal bedrooms opened. These were commensurate with the dining-room and the two sitting-rooms below. From this gallery there led a long narrow passage shut off from the rest of the house by a red-baize door, which led to a couple more guest-rooms and the servants' quarters.

Jack Singleton and I share the same flat in town, and we had sent down in the morning Franklyn and his wife, two old and valued servants, to get things ready at Trevor Major, and procure help from the village to look after the house, and Mrs Franklyn, with her stout comfortable face all wreathed in smiles, opened the door to us. She had had some previous experience of the 'comfortable quarters' which go with fishing, and had come down prepared for the worst, but found it all of the best. The kitchen-boiler was not furred; hot and cold water was laid on in the most convenient fashion, and could be obtained from taps that neither stuck nor leaked.

Her husband, it appeared, had gone into the village to buy a few necessaries, and she brought up tea for us, and then went upstairs to the two rooms over the dining-room and bigger sitting-room, which we had chosen for our bedrooms, to unpack. The doors of these were exactly opposite one another to right and left of the gallery, and Jack, who chose the bedroom above the sitting-room, had thus a smaller room, above the second sitting-room, unoccupied, next his and opening out from it.

We had a couple of hours' fishing before dinner, each of us catching three or four brace of trout, and came back in the dusk to the house. Franklyn had returned from the village from his errand, reported that he had got a woman to come in to do housework in the mornings, and mentioned that our arrival had seemed to arouse a good deal of interest.

The reason for this was obscure; he could only tell us that he was questioned a dozen times as to whether we really intended to live in the house, and his assurance that we did produced silence and a shaking of heads. But the countryfolk of Sussex are notable for their silence and chronic attitude of disapproval, and we put this down to local idiosyncrasy.

The evening was exquisitely warm, and after dinner we pulled out a couple of basket-chairs on to the gravel by the front door, and sat for an hour or so, while the night deepened in throbs of gathering darkness.

The moon was not risen and the ring of pines cut off much of the pale starlight, so that when we went in, allured by the shining of the lamp in the sitting-room, it was curiously dark for a clear night in May. And at the moment of stepping from the darkness into the cheerfulness of the lighted house, I had a sudden sensation, to which, during the next fortnight, I became almost accustomed, of there being something unseen and unheard and dreadful near me.

In spite of the warmth, I felt myself shiver, and concluded instantly that I had sat out-of-doors long enough, and without mentioning it to Jack, followed him into the smaller sitting-room in which we had scarcely yet set foot. It, like the hall, was oak-panelled, and in the panels hung some half-dozen of water-colour sketches, which we examined, idly at first, and then with growing interest, for they were executed with extraordinary finish and delicacy, and each represented some aspect of the house or garden.

Here you looked up the gap in the fir-trees into a crimson sunset; here the garden, trim and carefully tended, dozed beneath some languid summer noon; here an angry wreath of storm-cloud brooded over the meadow where the trout-stream ran grey and leaden below a threatening sky, while another, the most careful and arresting of all, was a study of the brick-kiln.

In this, alone of them all, was there a human figure; a man, dressed in grey, peered into the open door from which issued a fierce red glow. The figure was painted with miniature-like elaboration; the face was in profile, and represented a youngish man, clean-shaven, with a long aquiline nose and singularly square chin. The sketch was long and narrow in shape, and the chimney of the kiln appeared against a dark sky. From it there issued a thin stream of grey smoke.

Jack looked at this with attention.

'What a horrible picture,' he said, 'and how beautifully painted! I feel as if it meant something, as if it was a representation of something that happened, not a mere sketch. By Jove!—'

He broke off suddenly and went in turn to each of the other pictures.

'That's a queer thing,' he said. 'See if you notice what I mean.'

With the brick-kiln rather vividly impressed on my mind, it was not difficult to see what he had noticed. In each of the pictures appeared the brick-kiln, chimney and all, now seen faintly between trees, now in full view, and in each the chimney was smoking.

'And the odd part is that from the garden side, you can't really see the kiln at all,' observed Jack, 'it's hidden by the house, and yet the artist F. A., as I see by his signature, puts it in just the same.'

'What do you make of that?' I asked.

'Nothing. I suppose he had a fancy for brick-kilns. Let's have a game of picquet.'

A fortnight of our three weeks passed without incident, except that again and again the curious feeling of something dreadful being close at hand was present in my mind. In a way, as I said, I got used to it, but on the other hand the feeling itself seemed to gain in poignancy. Once just at the end of the fortnight I mentioned it to Jack.

'Odd you should speak of it,' he said, 'because I've felt the same. When do you feel it? Do you feel it now, for instance?'

We were again sitting out after dinner, and as he spoke I felt it with far greater intensity than ever before. And at the same moment the house-door, which had been closed, though probably not latched, swung gently open, letting out a shaft of light from the hall, and as gently swung to again, as if something had stealthily entered.

'Yes,' I said. 'I felt it then. I only feel it in the evening. It was rather bad that time.'

Jack was silent a moment.

'Funny thing the door opening and shutting like that,' he said. 'Let's go indoors.'

We got up and I remembered seeing at that moment that the windows of my bedroom were lit; Mrs Franklyn probably was making things ready for the night. Simultaneously, as we crossed the gravel, there came from just inside the house the sound of a hurried footstep on the stairs, and entering we found Mrs Franklyn in the hall, looking rather white and startled.

'Anything wrong?' I asked.

She took two or three quick breaths before she answered:

'No, sir,' she said, 'at least nothing that I can give an account of. I was tidying up in your room, and I thought you came in. But there was nobody, and it gave me a turn. I left my candle there; I must go up for it.'

I waited in the hall a moment, while she again ascended the stairs, and passed along the gallery to my room. At the door, which I could see was open, she paused, not entering.

'What is the matter?' I asked from below.

'I left the candle alight,' she said, 'and it's gone out.'

Jack laughed.

'And you left the door and window open,' said he.

'Yes, sir, but not a breath of wind is stirring,' said Mrs Franklyn, rather faintly.

This was true, and yet a few moments ago the heavy hall-door had swung open and back again. Jack ran upstairs.

'We'll brave the dark together, Mrs Franklyn,' he said.

He went into my room, and I heard the sound of a match struck. Then through the open door came the light of the rekindled candle and simultaneously I heard a bell ring in the servants' quarters. In a moment came steps, and Franklyn appeared.

'What bell was that?' I asked.

'Mr Jack's bedroom, sir,' he said.

I felt there was a marked atmosphere of nerves about for which there was really no adequate cause. All that had happened of a disturbing nature was that Mrs Franklyn had thought I had come into my bedroom, and had been startled by finding I had not. She had then left the candle in a draught, and it had been blown out. As for a bell ringing, that, even if it had happened, was a very innocuous proceeding.

'Mouse on a wire,' I said. 'Mr Jack is in my room this moment lighting Mrs Franklyn's candle for her.'

Jack came down at this juncture, and we went into the sitting-room. But Franklyn apparently was not satisfied, for we heard him in the room above us, which was Jack's bedroom, moving about with his slow and rather ponderous tread. Then his steps seemed to pass into the bedroom adjoining and we heard no more.

I remember feeling hugely sleepy that night, and went to bed earlier than usual, to pass rather a broken night with stretches of dreamless sleep interspersed with startled awakenings, in which I passed very suddenly into complete consciousness. Sometimes the house was absolutely still, and the only sound to be heard was the sighing of the night breeze outside in the pines, but sometimes the place seemed full of muffled movements and once I could have sworn that the handle of my door turned.

That required verification, and I lit my candle, but found that my ears must have played me false. Yet even as I stood there, I thought I heard steps just outside, and with a considerable qualm, I must confess, I opened the door and looked out. But the gallery was quite empty, and the house quite still. Then from Jack's room opposite I heard a sound that was somehow comforting, the snorts of the snorer, and I went back to bed and slept again, and when next I woke, morning was already breaking in red lines on the horizon, and the sense of trouble that had been with me ever since last evening had gone.

Heavy rain set in after lunch next day, and as I had arrears of letter-writing to do, and the water was soon both muddy and rising, I came home alone about five, leaving Jack still sanguine by the stream, and worked for a couple of hours sitting at a writing-table in the room overlooking the gravel at the front of the house, where hung the water-colours.

By seven I had finished, and just as I got up to light candles, since it was already dusk, I saw, as I thought, Jack's figure emerge from the bushes that bordered the path to the stream, on to the space in front of the house. Then instantaneously and with a sudden queer sinking of the heart quite unaccountable, I saw that it was not Jack at all, but a stranger.

He was only some six yards from the window, and after pausing there a moment he came close up to the window, so that his face nearly touched the glass, looking intently at me. In the light from the freshly kindled candles I could distinguish his features with great clearness, but though, as far as I knew, I had never seen him before, there was something familiar about both his face and figure. He appeared to smile at me, but the smile was one of inscrutable evil and malevolence, and immediately he walked on, straight towards the house door opposite him, and out of sight of the sitting-room window.

Now, little though I liked the look of the man, he was as I have said, familiar to my eye, and I went out into the hall, since he was clearly coming to the front-door, to open it to him and learn his business. So without waiting for him to ring, I opened it, feeling sure I should find him on the step. Instead, I looked out into the empty gravel-sweep, the heavy-falling rain, the thick dusk. And even as I looked, I felt something that I could not see push by me through the half-opened door and pass into the house. Then the stairs creaked, and a moment after a bell rang.

Franklyn is the quickest man to answer a bell I have ever seen, and next instant he passed me going upstairs. He tapped at Jack's door, entered and then came down again.

'Mr Jack still out, sir?' he asked.

'Yes. His bell ringing again?'

'Yes, sir,' said Franklyn, quite imperturbably.

I went back into the sitting-room, and soon Franklyn brought a lamp. He put it on the table above which hung the careful and curious picture of the brick-kiln, and then with a sudden horror I saw why the stranger on the gravel outside had been so familiar to me. In all respects he resembled the figure that peered into the kiln; it was more than a resemblance, it was an identity. And what had happened to this man who had inscrutably and evilly smiled at me? And what had pushed in through the half-closed door?

At that moment I saw the face of Fear; my mouth went dry, and I heard my heart leaping and cracking in my throat. That face was only turned on me for a moment, and then away again, but I knew it to be the genuine thing; not apprehension, not foreboding, not a feeling of being startled, but Fear, cold Fear. And then though nothing had occurred to assuage the Fear, it passed, and a certain sort of reason usurped—for so I must say—its place.

I had certainly seen somebody on the gravel outside the house; I had supposed he was going to the front-door. I had opened it, and found he had not come to the front-door. Or—and once again the terror resurged—had the invisible pushing thing been that which I had seen outside? And if so, what was it? And how came it that the face and figure of the man I had seen were the same as those which were so scrupulously painted in the picture of the brick-kiln?

I set myself to argue down the Fear for which there was no more foundation than this, this and the repetition of the ringing bell, and my belief is that I did so.

I told myself, till I believed it, that a man—a human man—had been walking across the gravel outside, and that he had not come to the front-door but had gone, as he might easily have done, up the drive into the high-road. I told myself that it was mere fancy that was the cause of the belief that Something had pushed in by me, and as for the ringing of the bell, I said to myself, as was true, that this had happened before. And I must ask the reader to believe also that I argued these things away, and looked no longer on the face of Fear itself. I was not comfortable, but I fell short of being terrified.

I sat down again by the window looking on to the gravel in front of the house, and finding another letter that asked, though it did not demand, an answer, proceeded to occupy myself with it.

Straight in front led the drive through the gap in the pines, and passed through the field where lay the brick-kiln. In a pause of page-turning I looked up and saw something unusual about it; at the same moment an unusual smell came to my nostril. What I saw was smoke coming out of the chimney of the kiln, what I smelt was the odour of roasting meat. The wind—such as there was—set from the kiln to the house. But as far as I knew the smell of roast meat probably came from the kitchen where dinner, so I supposed, was cooking. I had to tell myself this: I wanted reassurance, lest the face of Fear should look whitely on me again.

Then there came a crisp step on the gravel, a rattle at the front-door, and Jack came in.

'Good sport,' he said, 'you gave up too soon.'

And he went straight to the table above which hung the picture of the man at the brick-kiln, and looked at it. Then there was silence; and eventually I spoke, for I wanted to know one thing.

'Seen anybody?' I asked.

'Yes. Why do you ask?'

'Because I have also; the man in that picture.'

Jack came and sat down near me.

'It's a ghost, you know,' he said. 'He came down to the river about dusk and stood near me for an hour. At first I thought he was—was real, and I warned him that he had better stand farther off if he didn't want to be hooked. And then it struck me he wasn't real, and I cast, well, right through him, and about seven he walked up towards the house.'

'Were you frightened?'

'No. It was so tremendously interesting. So you saw him here too. Whereabouts?'

'Just outside. I think he is in the house now.'

Jack looked round.

'Did you see him come in?' he asked.

'No, but I felt him. There's another queer thing too; the chimney of the brick-kiln is smoking.'

Jack looked out of the window. It was nearly dark, but the wreathing smoke could just be seen.

'So it is,' he said, 'fat, greasy smoke. I think I'll go up and see what's on. Come too?'

'I think not,' I said.

'Are you frightened? It isn't worth while. Besides, it is so tremendously interesting.'

Jack came back from his little expedition still interested. He had found nothing stirring at the kiln, but though it was then nearly dark the interior was faintly luminous, and against the black of the sky he could see a wisp of thick white smoke floating northwards. But for the rest of the evening we neither heard nor saw anything of abnormal import, and the next day ran a course of undisturbed hours. Then suddenly a hellish activity was manifested.

That night, while I was undressing for bed, I heard a bell ring furiously, and I thought I heard a shout also. I guessed where the ring came from, since Franklyn and his wife had long ago gone to bed, and went straight to Jack's room. But as I tapped at the door I heard his voice from inside calling loud to me. 'Take care,' it said, 'he's close to the door.'

A sudden qualm of blank fear took hold of me, but mastering it as best I could, I opened the door to enter, and once again something pushed softly by me, though I saw nothing.

Jack was standing by his bed, half-undressed. I saw him wipe his forehead with the back of his hand.

'He's been here again,' he said. 'I was standing just here, a minute ago, when I found him close by me. He came out of the inner room, I think. Did you see what he had in his hand?'

'I saw nothing.'

'It was a knife; a great long carving knife. Do you mind my sleeping on the sofa in your room tonight? I got an awful turn then. There was another thing too. All round the edge of his clothes, at his collar and at his wrists, there were little flames playing, little white licking flames.'

But next day, again, we neither heard nor saw anything, nor that night did the sense of that dreadful presence in the house come to us. And then came the last day.

We had been out till it was dark, and as I said, had a wonderful day among the fish. On reaching home we sat together in the sitting-room, when suddenly from overhead came a tread of feet, a violent pealing of the bell, and the moment after yell after yell as of someone in mortal agony. The thought occurred to both of us that this might be Mrs Franklyn in terror of some fearful sight, and together we rushed up and sprang into Jack's bedroom.

The doorway into the room beyond was open, and just inside it we saw the man bending over some dark huddled object.

Though the room was dark we could see him perfectly, for a light stale and impure seemed to come from him. He had again a long knife in his hand, and as we entered he was wiping it on the mass that lay at his feet. Then he took it up, and we saw what it was, a woman with head nearly severed. But it was not Mrs Franklyn.

And then the whole thing vanished, and we were standing looking into a dark and empty room. We went downstairs without a word, and it was not till we were both in the sitting-room below that Jack spoke.

'And he takes her to the brick-kiln,' he said rather unsteadily. 'I say, have you had enough of this house? I have. There is hell in it.'

About a week later Jack put into my hand a guide-book to Sussex open at the description of Trevor Major, and I read:

'Just outside the village stands the picturesque manor house, once the home of the artist and notorious murderer, Francis Adam. It was here he killed his wife, in a fit, it is believed, of groundless jealousy, cutting her throat and disposing of her remains by burning them in a brick-kiln. Certain charred fragments found six months afterwards led to his arrest and execution.'

So I prefer to leave the house with the brick-kiln and the pictures signed F. A. to others.

The Monkeys

RUSKIN BOND

I couldn't be sure, next morning, if I had been dreaming or if I had really heard dogs barking in the night and had seen them scampering about on the hillside below the cottage. There had been a Golden Cocker, a Retriever, a Peke, a Dachshund, a black Labrador, and one or two nondescripts. They had woken me with their barking shortly after midnight, and made so much noise that I got out of bed and looked out of the open window. I saw them quite plainly in the moonlight, five or six dogs rushing excitedly through the bracken and long monsoon grass.

It was only because there had been so many breeds among the dogs that I felt a little confused. I had been in the cottage only a week, and I was already on nodding or speaking terms with most of my neighbours.

Colonel Fanshawe, retired from the Indian Army, was my immediate neighbour. He did keep a Cocker, but it was black. The elderly Anglo-Indian spinsters who lived beyond the deodars kept only cats. (Though why cats should be the prerogative of spinsters, I have never been able to understand.) The milkman kept a couple of mongrels. And the Punjabi industrialist who had bought a former prince's palace—without ever occupying it—left the property in charge of a watchman who kept a huge Tibetan mastiff.

None of these dogs looked like the ones I had seen in the night.

'Does anyone here keep a Retriever?' I asked Colonel Fanshawe, when I met him taking his evening walk.

'No one that I know of,' he said, and he gave me a swift, penetrating look from under his bushy eyebrows. 'Why, have you seen one around?'

'No, I just wondered. There are a lot of dogs in the area, aren't there?'

'Oh, yes. Nearly everyone keeps a dog here. Of course every now and then a panther carries one off. Lost a lovely little terrier myself, only last winter.'

Colonel Fanshawe, tall and red-faced, seemed to be waiting for me to tell him something more—or was he just taking time to recover his breath after a stiff uphill climb?

That night I heard the dogs again. I went to the window and looked out. The moon was at the full, silvering the leaves of the oak trees.

The dogs were looking up into the trees, and barking. But I could see nothing in the trees, not even an owl.

I gave a shout, and the dogs disappeared into the forest.

Colonel Fanshawe looked at me expectantly when I met him the following day. He knew something about those dogs, of that I was certain; but he was waiting to hear what I had to say. I decided to oblige him.

'I saw at least six dogs in the middle of the night,' I said. 'A Cocker, a Retriever, a Peke, a Dachshund, and two mongrels. Now, Colonel, I'm sure you must know whose they are.'

The Colonel was delighted. I could tell by the way his eyes glinted that he was going to enjoy himself at my expense.

'You've been seeing Miss Fairchild's dogs,' he said with smug satisfaction.

'Oh, and where does she live?'

'She doesn't, my boy. Died fifteen years ago.'

'Then what are her dogs doing here?'

'Looking for monkeys,' said the Colonel. And he stood back to watch my reactions.

'I'm afraid I don't understand,' I said.

'Let me put it this way,' said the Colonel. 'Do you believe in ghosts?'

'I've never seen any,' I said.

'But you have, my boy, you have. Miss Fairchild's dogs died years ago—a Cocker, a Retriever, a Dachshund, a Peke, and two mongrels. They were buried on a little knoll under the oaks. Nothing odd about their deaths, mind you. They were all quite old, and didn't survive their mistress very long. Neighbours looked after them until they died.'

'And Miss Fairchild lived in the cottage where I stay? Was she young?'

'She was in her mid-forties, an athletic sort of woman, fond of the outdoors. Didn't care much for men. I thought you knew about her.'

'No, I haven't been here very long, you know. But what was it you said about monkeys? Why were the dogs looking for monkeys?'

'Ah, that's the interesting part of the story. Have you seen the *langur* monkeys that sometimes come to eat oak leaves?'

'No.'

'You will, sooner or later. There has always been a band of them roaming these forests. They're quite harmless really, except that they'll ruin a garden if given half a chance . . . Well, Miss Fairchild fairly loathed those monkeys. She was very keen on her dahlias—grew some prize specimens—but the monkeys would come at night, dig up the plants, and eat the dahlia bulbs. Apparently they found the bulbs much to their liking. Miss Fairchild would be furious. People who are passionately fond of gardening often go off balance when their best plants are ruined—that's only human, I suppose. Miss Fairchild set her dogs at the monkeys, whenever she could, even if it was in the middle of the night. But the monkeys simply took to the trees and left the dogs barking.'

'Then one day—or rather, one night—Miss Fairchild took desperate measures. She borrowed a shotgun, and sat up near a window. And when the monkeys arrived, she shot one of them dead.'

The Colonel paused and looked out over the oak trees which were shimmering in the warm afternoon sun.

'She shouldn't have done that,' he said. 'Never shoot a monkey. It's not only that they're sacred to Hindus—but they are rather human, you know. Well, I must be getting on. Good-day!' And the Colonel, having ended his story rather abruptly, set off at a brisk pace through the deodars.

I didn't hear the dogs that night. But next day I saw the monkeys—the real ones, not ghosts. There were about twenty of them, young and old, sitting in the trees munching oak leaves. They didn't pay much attention to me, and I watched them for some time.

They were handsome creatures, their fur a silver-grey, their tails long and sinuous. They leapt gracefully from tree to tree, and were very polite and dignified in their behaviour towards each other—unlike the bold, rather crude red monkeys of the plains. Some of the younger ones scampered about on the hillside, playing and wrestling with each other like schoolboys.

There were no dogs to molest them—and no dahlias to tempt them into the garden.

But that night, I heard the dogs again. They were barking more furiously than ever.

'Well, I'm not getting up for them this time,' I mumbled, and pulled the blankets over my ears.

But the barking grew louder, and was joined by other sounds, a squealing and a scuffling.

Then suddenly the piercing shriek of a woman rang through the forest. It was an unearthly sound, and it made my hair stand up.

I leapt out of bed and dashed to the window.

A woman was lying on the ground, and three or four huge monkeys were on top of her, biting her arms and pulling at her throat. The dogs were yelping and trying to drag the monkeys off, but they were being harried from behind by others. The woman gave another bloodcurdling shriek, and I dashed back into the room, grabbed hold of a small axe, and ran into the garden.

But everyone—dogs, monkeys and shrieking woman—had disappeared, and I stood alone on the hillside in my pyjamas, clutching an axe and feeling very foolish.

The Colonel greeted me effusively the following day.

'Still seeing those dogs?' he asked in a bantering tone.

'I've seen the monkeys too,' I said.

'Oh, yes, they've come around again. But they're real enough, and quite harmless.'

'I know—but I saw them last night with the dogs.'

'Oh, did you really? That's strange, very strange.'

The Colonel tried to avoid my eye, but I hadn't quite finished with him.

'Colonel,' I said. 'You never did get around to telling me how Miss Fairchild died.'

'Oh, didn't I? Must have slipped my memory. I'm getting old, don't remember people as well as I used to. But of course I remember about Miss Fairchild, poor lady. The monkeys killed her. Didn't you know? They simply tore her to pieces . . . '

His voice trailed off, and he looked thoughtfully at a caterpillar that was making its way up his walking stick.

'She shouldn't have shot one of them,' he said. 'Never shoot a monkey—they're rather human, you know . . . '

Room at the Inn

SYDNEY J. BOUNDS

S unset laid a sheen of blood over bleak and desolate moorland. Wind came in great gusts, blowing the two cycles across the empty road, carrying with it an icy lash of rain.

Jane Black and her younger sister, Penny, on a cycling tour of Cornwall, were somewhere on Bodmin Moor.

'This,' Jane said a trifle grimly, 'is no longer funny. We'll have to stop and cape up.'

The two girls, both in jeans and wind-cheaters, got off their bikes, unrolled yellow oilskins and put them on. When they started off again, the wind howled in fury and threatened to blow them right off the road. The sun had almost gone and dark clouds piled high in a granite sky.

'It's a pity we ever left the main road,' Penny gasped, head down and wrestling with her machine to keep it upright. 'Any idea where we are?' Her words were whipped away.

'What was that?' Jane shouted.

'I said . . . oh, never mind. . . . '

The sky split and rain deluged down.

'We'll never get a tent up in this,' Jane said. 'Have to press on till we find some kind of shelter.'

They were miles from any village and there wasn't a house, or even a car, in sight. Jane had the eerie notion they were lone survivors, cut off from the rest of the world.

Heads down, braced against the wind, they cycled through drenching rain.

'A light!' Penny pointed, and fell off her bike.

Jane braked. 'You all right?'

'Of course I am. . . . Look over there.'

To the left, standing back from the road and on the crest of a tor, was the shadow of a house with a yellow light burning.

'Come on,' Jane said briskly. 'No one can refuse us shelter tonight.'

They pushed their bikes towards the building and entered a yard; outhouses cut off some of the wind. They put their bikes out of the rain, unstrapped their saddle-bags and made a run for the door of the house. A dark sign swung creakily overhead.

'It's an inn—and they won't hear us in this lot.'

Boldly, Jane pushed open the door and stepped into a stone-flagged passage. 'Anyone home?'

Just off the passage was a small bar with bottles and casks; an oil lamp burned here. The rest of the house was in darkness, silent. Her voice echoed in emptiness.

A girl, not much older than Jane, appeared out of the gloom. Her skirt touched the floor; a shawl draped her shoulders and her hair was done in ringlets.

'My uncle's away just now—'

'Can you put us up for the night?' Jane said quickly. 'We're wet through.'

'And cold. And hungry,' Penny added.

The girl looked doubtful. 'My uncle won't like—' She hesitated. 'But I can't turn you away in this weather. Yes, you can stay the night.'

'Thanks. We're cycling round Cornwall and—'

'Cycling?' The girl looked blank, then said: 'I'm called Jessica. Come through to the kitchen and get out of those wet things.'

Jane and Penny followed her along a dark passage to a kitchen with an oil lamp and stove. They stripped and Jessica brought them each a blanket to wrap round them.

They sat at a plain wood table, on hard-backed chairs, and Jessica poured hot soup into bowls. 'I'm afraid there isn't much else,' she said. 'We don't get many people stopping here.'

'Soup's lovely,' Penny said. 'I feel warm and sleepy already.'

Jessica lit a candle. 'This way, then.'

They went up a flight of wooden stairs, the boards creaking, to another passage with doors leading off. Jessica opened one. 'In here.'

Penny saw that the next door, the last in the passage, was closed and padlocked. 'What's in there?' she asked, curious.

Jane said quickly: 'It's none of your business,' and pushed her into their bedroom.

The walls were plaster, the boards bare. There was an old-fashioned double-bed and grime on a window which looked as if it had not been opened for years.

Jessica handed Jane a large iron key, one end of which was cross-shaped. 'Lock your door,' she said solemnly. 'Don't open it no matter what you hear tonight.'

'What are we likely to hear?' Penny asked promptly.

But Jessica shut the door on them.

Jane put the key in the lock and turned it, felt the bed. 'Damp—lucky we've got our sleeping bags.'

'Pretty primitive, the whole place,' Penny sniffed. 'And Jessica—I think she's weird.'

Snug in her sleeping bag, the last thing Jane heard was the drumming of the rain.

She woke reluctantly, becoming slowly aware of a hand shaking her shoulder and Penny's urgent voice: 'Something funny's going on.'

'Not funny . . . go back to sleep.'

'Oh, wake up, Jane! Please!'

Jane realized that the wind and rain had stopped, that drunken singing rose from below. Penny pushed a wristwatch before her eyes.

'It's three in the morning—and just look out here!' Penny crossed to the window and pointed down.

Jane wriggled out of her sleeping bag and padded over bare boards. A full moon hung in the sky. Wheels rumbled noisily as a train of horse-drawn wagons pulled into the yard below. She heard coarse laughter follow some of the words.

The men with the wagons were dressed in jerseys and dark, flaring trousers; some had knives and one an eye-patch. Their leader, big and bearded, cursed them.

'A villainous lot,' Penny said. 'What d'you suppose they're doing?'

Large boxes were dragged from the wagons and let fall, heavily, on the ground. Staggering, the men carried them, one by one, through the inn porch.

Presently, there came a bumping sound as the boxes were hauled up the stairs. Boots echoed in the passage. Wood scraped as the boxes were dragged past the girls' bedroom. Iron rattled.

'The locked room next door,' Penny whispered excitedly. 'Do you think they're smugglers?'

'Don't be silly,' Jane said, 'and keep quiet.' Remembering what Jessica had told them, she added: 'Anyway, it's nothing to do with us.'

Heavy footfalls came along the passage, stopped outside their door. A hand rattled the door-handle and a harsh voice demanded: 'Anyone in here?'

Jane clamped a hand over Penny's mouth, and Jessica's voice sounded in the passage:

'Two guests, caught by the weather. They're locked in, and drugged.'

The man laughed unpleasantly. 'Lucky for them. Show their noses tonight and they'll get their throats cut.'

Penny shivered. 'I don't like this,' she whispered, 'I'm scared.'

There were more scraping noises as boxes were shoved into the room next door. The men made several trips, then the door was locked again.

Boots trod the bare boards, going downstairs. The last pair paused outside the bedroom . . .

'Don't even breathe,' Jane whispered, hugging her sister.

Finally, the last man moved away and the passage was silent again. There came a creak of harness and a clop-clop of hooves.

Cautiously, Jane peered from the window. She saw the wagon train leaving the inn, heading out across the moonlit moor.

'It's all right, Penny, they've gone now.'

The inn seemed strangely quiet as the two girls wriggled deep into sleeping bags and, despite a sense of unease, both were soon asleep.

Jane woke with the sun in her eyes. Her hip and shoulder ached from lying on hard and uneven ground. Ground? Suddenly she remembered where she was and sat up. There was no roof, no walls—only a few flagstones scattered around. She lay on the ground with the moor stretching away into the distance.

Penny was still asleep, curled up in her sleeping bag, and their bikes lay propped against a stone some yards away.

Jane, realizing with a chill that she lay in the ruins of the inn, pinched herself. She was sure she was awake. . . .

She shook Penny and her sister woke suddenly. 'They haven't come back—?'

Penny looked at the ruins in disbelief. 'It couldn't have been a dream. It was too real!'

'Our clothes . . . lucky there's no one about to see us.'

Jane was sure their clothes lay where the kitchen stove would have been. Abruptly, she said, 'Come on, let's get away from here.'

They dressed quickly and rolled their sleeping bags, and as Jane shook hers, something metallic fell out. A rusty iron key with a cross at one end.

The Crown Derby Plate

MARJORIE BOWEN

Martha Pym said that she had never seen a ghost and that she would very much like to do so, 'particularly at Christmas, for you can laugh as you like, that is the correct time to see a ghost.'

'I don't suppose you ever will,' replied her cousin Mabel comfortably, while her cousin Clara shuddered and said that she hoped they would change the subject for she disliked even to think of such things.

The three elderly, cheerful women sat round a big fire, cosy and content after a day of pleasant activities; Martha was the guest of the other two, who owned the handsome, convenient country house; she always came to spend her Christmas with the Wyntons and found the leisurely country life delightful after the bustling round of London, for Martha managed an antique shop of the better sort and worked extremely hard. She was, however, still full of zest for work or pleasure, though sixty years old, and looked backwards and forwards to a succession of delightful days.

The other two, Mabel and Clara, led quieter but none the less agreeable lives; they had more money and fewer interests, but nevertheless enjoyed themselves very well.

'Talking of ghosts,' said Mabel, 'I wonder how that old woman at Hartleys is getting on, for Hartleys, you know, is supposed to be haunted.'

'Yes, I know,' smiled Miss Pym, 'but all the years that we have known of the place we have never heard anything definite, have we?'

'No,' put in Clara, 'but there *is* that persistent rumour that the house is uncanny, and for myself, *nothing* would induce me to live there!'

'It is certainly very lonely and dreary down there on the marshes,' conceded Mabel. 'But as for the ghost—you never hear *what* it is supposed to be even.'

'Who has taken it?' asked Miss Pym, remembering Hartleys as very desolate indeed, and long shut up.

'A Miss Lefain, an eccentric old creature—I think you met her here once, two years ago—'

'I believe that I did, but I don't recall her at all.'

'We have not seen her since, Hartleys is so un-get-at-able and she didn't seem to want visitors. She collects china, Martha, so really you ought to go and see her and talk "shop".'

With the word 'china' some curious associations came into the mind of Martha Pym; she was silent while she strove to put them together, and after a second or two they all fitted together into a very clear picture.

She remembered that thirty years ago—yes, it must be thirty years ago, when, as a young woman, she had put all her capital into the antique business, and had been staying with her cousins (her aunt had then been alive) that she had driven across the marsh to Hartleys, where there was an auction sale; all the details of this she had completely forgotten, but she could recall quite clearly purchasing a set of gorgeous china which was still one of her proud delights, a perfect set of Crown Derby save that one plate was missing.

'How odd,' she remarked, 'that this Miss Lefain should collect china too, for it was at Hartleys that I purchased my dear old Derby service—I've never been able to match that plate—'

'A plate was missing? I seem to remember,' said Clara. 'Didn't they say that it must be in the house somewhere and that it should be looked for?'

'I believe they did, but of course I never heard any more and that missing plate has annoyed me ever since. Who had Hartleys?'

'An old connoisseur, Sir James Sewell; I believe he was some relation to this Miss Lefain, but I don't know—'

'I wonder if she has found the plate,' mused Miss Pym. 'I expect she has turned out and ransacked the whole place—'

'Why not trot over and ask?' suggested Mabel. 'It's not much use to her, if she has found it, one odd plate.'

'Don't be silly,' said Clara. 'Fancy going over the marshes, this weather, to ask about a plate missed all those years ago. I'm sure Martha wouldn't think of it—'

But Martha did think of it; she was rather fascinated by the idea; how queer and pleasant it would be if, after all these years, nearly a lifetime, she should find the Crown Derby plate, the loss of which had always irked her! And this hope did not seem so altogether fantastical, it was quite likely that old Miss Lefain, poking about in the ancient house, had found the missing piece.

And, of course, if she had, being a fellow-collector, she would be quite willing to part with it to complete the set.

Her cousin endeavoured to dissuade her; Miss Lefain, she declared, was a recluse, an odd creature who might greatly resent such a visit and such a request.

'Well, if she does I can but come away again,' smiled Miss Pym. 'I suppose she can't bite my head off, and I rather like meeting these curious types—we've got a love for old china in common, anyhow.'

'It seems so silly to think of it—after all these years—a plate!'

'A Crown Derby plate,' corrected Miss Pym. 'It is certainly strange that I didn't think of it before, but now that I have got it into my head I can't get it out. Besides,' she added hopefully, 'I might see the ghost.'

So full, however, were the days with pleasant local engagements that Miss Pym had no immediate chance of putting her scheme into practice; but she did not relinquish it, and she asked several different people what they knew about Hartleys and Miss Lefain.

And no one knew anything save that the house was supposed to be haunted and the owner 'cracky'.

'Is there a story?' asked Miss Pym, who associated ghosts with neat tales into which they fitted as exactly as nuts into shells.

But she was always told—'Oh, no, there isn't a story, no one knows anything about the place, don't know how the idea got about; old Sewell was half-crazy, I believe, he was buried in the garden and that gives a house a nasty name—'

'Very unpleasant,' said Martha Pym, undisturbed.

This ghost seemed too elusive for her to track down; she would have to be content if she could recover the Crown Derby plate; for that at least she was determined to make a try and also to satisfy that faint tingling of curiosity roused in her by this talk about Hartleys and the remembrance of that day, so long ago, when she had gone to the auction sale at the lonely old house.

So the first free afternoon, while Mabel and Clara were comfortably taking their afternoon repose, Martha Pym, who was of a more lively habit, got out her little governess cart and dashed away across the Essex flats.

She had taken minute directions with her, but she had soon lost her way.

Under the wintry sky, which looked as grey and hard as metal, the marshes stretched bleakly to the horizon, the olive-brown broken reeds were harsh as scars on the saffron-tinted bogs, where the sluggish waters that rose so high in winter were filmed over with the first stillness of a frost; the air was cold but not keen, everything was damp; the faintest of mists blurred the black outlines of trees that rose stark from the ridges above the stagnant dykes; the flooded fields were haunted by black birds and white birds, gulls and crows whining above the long ditch grass and wintry wastes.

Miss Pym stopped the little horse and surveyed this spectral scene, which had a certain relish about it to one sure to return to a homely village, a cheerful house and good company.

A withered and bleached old man, in colour like the dun landscape, came along the road between the sparse alders.

Miss Pym, buttoning up her coat, asked the way to Hartleys as he passed her; he told her, straight on, and she proceeded, straight indeed along the road that went with undeviating length across the marshes.

'Of course,' thought Miss Pym, 'if you live in a place like this, you are bound to invent ghosts.'

The house sprang up suddenly on a knoll ringed with rotting trees, encompassed by an old brick wall that the perpetual damp had overrun with lichen, blue, green, white colours of decay.

Hartleys, no doubt, there was no other residence of human being in sight in all the wide expanse; besides, she could remember it, surely, after all this time, the sharp rising out of the marsh, the colony of tall trees, but then fields and trees had been green and bright—there had been no water on the flats, it had been summer time.

'She certainly,' thought Miss Pym, 'must be crazy to live here. And I rather doubt if I shall get my plate.'

She fastened up the good little horse by the garden gate which stood negligently ajar and entered; the garden itself was so neglected that it was quite surprising to see a trim appearance in the house, curtains at the window and a polish on the brass door knocker, which must have been recently rubbed there, considering the taint in the sea damp which rusted everything.

It was a square-built, substantial house with 'nothing wrong with it but the situation', Miss Pym decided, though it was not very attractive, being built of that drab plastered stone so popular a hundred years ago, with flat windows and door, while one side was gloomily shaded by a large evergreen tree of the cypress variety which gave a blackish tinge to that portion of the garden.

There was no pretence at flower-beds nor any manner of cultivation in this garden where a few rank weeds and straggling bushes matted together above the dead grass; on the enclosing

wall which appeared to have been built high as protection against the ceaseless winds that swung along the flats were the remains of fruit trees; their crucified branches, rotting under the great nails that held them up, looked like the skeletons of those who had died in torment.

Miss Pym took in these noxious details as she knocked firmly at the door; they did not depress her; she merely felt extremely sorry for anyone who could live in such a place.

She noticed, at the far end of the garden, in the corner of the wall, a headstone showing above the sodden colourless grass, and remembered what she had been told about the old antiquary being buried there, in the grounds of Hartleys.

As the knock had no effect she stepped back and looked at the house; it was certainly inhabited—with those neat windows, white curtains and drab blinds all pulled to precisely the same level.

And when she brought her glance back to the door she saw that it had been opened and that someone, considerably obscured by the darkness of the passage, was looking at her intently.

'Good afternoon,' said Miss Pym cheerfully. 'I just thought that I would call to see Miss Lefain—it is Miss Lefain, isn't it?'

'It's my house,' was the querulous reply.

Martha Pym had hardly expected to find any servants here, though the old lady must, she thought, work pretty hard to keep the house so clean and tidy as it appeared to be.

'Of course,' she replied. 'May I come in? I'm Martha Pym, staying with the Wyntons, I met you there—'

'Do come in,' was the faint reply. 'I get so few people to visit me, I'm really very lonely.'

'I don't wonder,' thought Miss Pym; but she had resolved to take no notice of any eccentricity on the part of her hostess, and so she entered the house with her usual agreeable candour and courtesy.

The passage was badly lit, but she was able to get a fair idea of Miss Lefain; her first impression was that this poor creature was most dreadfully old, older than any human being had the right to be, why, she felt young in comparison—so faded, feeble and pallid was Miss Lefain.

She was also monstrously fat; her gross, flaccid figure was shapeless and she wore a badly cut, full dress of no colour at all, but stained with earth and damp where Miss Pym supposed she had been doing futile gardening; this gown was doubtless designed to disguise her stoutness, but had been so carelessly pulled about that it only added to it, being rucked and rolled 'all over the place' as Miss Pym put it to herself.

Another ridiculous touch about the appearance of the poor old lady was her short hair; decrepit as she was, and lonely as she lived she had actually had her scanty relics of white hair cropped round her shaking head.

'Dear me, dear me,' she said in her thin treble voice. 'How very kind of you to come. I suppose you prefer the parlour? I generally sit in the garden.'

'The garden? But not in this weather?'

'I get used to the weather. You've no idea how used one gets to the weather.'

'I suppose so,' conceded Miss Pym doubtfully. 'You don't live here quite alone, do you?'

'Quite alone, lately. I had a little company, but she was taken away, I'm sure I don't know where. I haven't been able to find a trace of her anywhere,' replied the old lady peevishly.

'Some wretched companion that couldn't stick it, I suppose,' thought Miss Pym. 'Well, I don't wonder—but someone ought to be here to look after her.'

They went into the parlour, which, the visitor was dismayed to see, was without a fire but otherwise well kept.

And there, on dozens of shelves, was a choice array of china at which Martha Pym's eyes glistened.

'Aha!' cried Miss Lefain. 'I see you've noticed my treasures! Don't you envy me? Don't you wish that you had some of those pieces?'

Martha Pym certainly did and she looked eagerly and greedily round the walls, tables and cabinets while the old woman followed her with little thin squeals of pleasure.

It was a beautiful little collection, most choicely and elegantly arranged, and Martha thought it marvellous that this feeble ancient creature should be able to keep it in such precise order as well as doing her own housework.

'Do you really do everything yourself here and live quite alone?' she asked, and she shivered even in her thick coat and wished that Miss Lefain's energy had risen to a fire, but then probably she lived in the kitchen, as these lonely eccentrics often did.

'There was someone,' answered Miss Lefain cunningly, 'but I had to send her away. I told you she's gone, I can't find her, and I am so glad. Of course,' she added wistfully, 'it leaves me very lonely, but then I couldn't stand her impertinence any longer. She used to say that it was *her* house and her collection of china! Would you believe it? She used to try to chase me away from looking at my own things!'

'How very disagreeable,' said Miss Pym, wondering which of the two women had been crazy. 'But hadn't you better get someone else?'

'Oh no,' was the jealous answer. 'I would rather be alone with my things, I daren't leave the house for fear someone takes them away— there was a dreadful time once when an auction sale was held here—'

'Were you here then?' asked Miss Pym; but indeed she looked old enough to have been anywhere.

'Yes, of course,' Miss Lefain replied rather peevishly, and Miss Pym decided that she must be a relation of old Sir James Sewell. Clara and Mabel had been very foggy about it all. 'I was very busy hiding all the china—but one set they got—a Crown Derby tea service—'

'With one plate missing!' cried Martha Pym. 'I bought it, and do you know, I was wondering if you'd found it—'

'I hid it,' piped Miss Lefain.

'Oh, you did, did you? Well, that's rather funny behaviour. Why did you hide the stuff away instead of buying it?'

'How could I buy what was mine?'

'Old Sir James left it to you, then?' asked Martha Pym, feeling very muddled.

'*She* bought a lot more,' squeaked Miss Lefain, but Martha Pym tried to keep her to the point.

'If you've got the plate,' she insisted, 'you might let me have it—I'll pay quite handsomely, it would be so pleasant to have it after all these years.'

'Money is no use to me,' said Miss Lefain mournfully. 'Not a bit of use. I can't leave the house or the garden.'

'Well, you have to live, I suppose,' replied Martha Pym cheerfully. 'And, do you know, I'm afraid you are getting rather morbid and dull, living here all alone—you really ought to have a fire—why, it's just on Christmas and very damp.'

'I haven't felt the cold for a long time,' replied the other; she seated herself with a sigh on one of the horsehair chairs and Miss Pym noticed with a start that her feet were covered only by a pair of white stockings; 'one of those nasty health fiends,' thought Miss Pym, 'but she doesn't look too well for all that.'

'So you don't think that you could let me have the plate?' she asked briskly, walking up and down, for the dark, neat, clean parlour was very cold indeed, and she thought that she couldn't stand this much longer; as there seemed no sign of tea or anything pleasant and comfortable she had really better go.

'I might let you have it,' sighed Miss Lefain, 'since you've been so kind as to pay me a visit. After all, one plate isn't much use, is it?'

'Of course not, I wonder you troubled to hide it—'

'I couldn't *bear*,' wailed the other, 'so see the things going out of the house!'

Martha Pym couldn't stop to go into all this; it was quite clear that the old lady was very eccentric indeed and that nothing very much could be done with her; no wonder that she had 'dropped out' of everything and that no one ever saw her or knew anything about her, though Miss Pym felt that some effort ought really to be made to save her from herself.

'Wouldn't you like a run in my little governess cart?' she suggested. 'We might go to tea with the Wyntons on the way back, they'd be delighted to see you, and I really think that you do want taking out of yourself.'

'I was taken out of myself some time ago,' replied Miss Lefain. 'I really was, and I couldn't leave my things—though,' she added with pathetic gratitude, 'it is very, very kind of you—'

'Your things would be quite safe, I'm sure,' said Martha Pym, humouring her. 'Who ever would come up here, this hour of a winter's day?'

'They do, oh, they do! And *she* might come back, prying and nosing and saying that it was all hers, all my beautiful china, hers!'

Miss Lefain squealed in her agitation and rising up ran round the wall fingering with flaccid yellow hands the brilliant glossy pieces on the shelves.

'Well, then, I'm afraid that I must go, they'll be expecting me, and it's quite a long ride; perhaps some other time you'll come and see us?'

'Oh, must you go?' quavered Miss Lefain dolefully. 'I do like a little company now and then and I trusted you from the first—the others, when they do come, are always after my things and I have to frighten them away!'

'Frighten them away!' replied Martha Pym. 'However do you do that?'

'It doesn't seem difficult, people are so easily frightened, aren't they?'

Miss Pym suddenly remembered that Hartleys had the reputation of being haunted—perhaps the queer old thing played on that; the lonely house with the grave in the garden was dreary enough around which to create a legend.

'I suppose you've never seen a ghost?' she asked pleasantly. 'I'd rather like to see one, you know—'

'There is no one here but myself,' said Miss Lefain.

'So you've never seen anything? I thought it must be all nonsense. Still, I do think it rather melancholy for you to live here all alone—'

Miss Lefain sighed:

'Yes, it's very lonely. Do stay and talk to me a little longer.' Her whistling voice dropped cunningly. 'And I'll give you the Crown Derby plate!'

'Are you sure you've really got it?' Miss Pym asked.

'I'll show you.'

Fat and waddling as she was, she seemed to move very lightly as she slipped in front of Miss Pym and conducted her from the room, going slowly up the stairs—such a gross odd figure in that clumsy dress with the fringe of white hair hanging on to her shoulders.

The upstairs of the house was as neat as the parlour, everything well in its place; but there was no sign of occupancy; the beds were covered with dust sheets, there were no lamps or fires set ready. 'I suppose,' said Miss Pym to herself, 'she doesn't care to show me where she really lives.'

But as they passed from one room to another, she could not help saying:

'Where do *you* live, Miss Lefain?'

'Mostly in the garden,' said the other.

Miss Pym thought of those horrible health huts that some people indulged in.

'Well, sooner you than me,' she replied cheerfully.

In the most distant room of all, a dark, tiny closet, Miss Lefain opened a deep cupboard and brought out a Crown Derby plate which her guest received with a spasm of joy, for it was actually that missing from her cherished set.

'It's very good of you,' she said in delight. 'Won't you take something for it, or let me do something for you?'

'You might come and see me again,' replied Miss Lefain wistfully.

'Oh yes, of course I should like to come and see you again.'

But now that she had got what she had really come for, the plate, Martha Pym wanted to be gone; it was really very dismal and depressing in the house and she began to notice a fearful smell— the place had been shut up too long, there was something damp rotting somewhere, in this horrid little dark closet no doubt.

'I really must be going,' she said hurriedly.

Miss Lefain turned as if to cling to her, but Martha Pym moved quickly away.

'Dear me,' wailed the old lady. 'Why are you in such haste?'

'There's—a smell,' murmured Miss Pym rather faintly.

She found herself hastening down the stairs, with Miss Lefain complaining behind her.

'How peculiar people are—*she* used to talk of a smell—'

'Well, you must notice it yourself.'

Miss Pym was in the hall; the old woman had not followed her, but stood in the semi-darkness at the head of the stairs, a pale shapeless figure.

Martha Pym hated to be rude and ungrateful but she could not stay another moment; she hurried away and was in her cart in a moment—really—that smell—

'Goodbye!' she called out with false cheerfulness, 'and thank you so much!'

There was no answer from the house.

Miss Pym drove on; she was rather upset and took another way than that by which she had come, a way that led past a little house raised above the marsh; she was glad to think that the poor old creature at Hartleys had such near neighbours, and she reined up the horse, dubious as to whether she should call someone and tell them that poor old Miss Lefain really wanted a little looking after, alone in a house like that, and plainly not quite right in her head.

A young woman, attracted by the sound of the governess cart, came to the door of the house and seeing Miss Pym called out, asking if she wanted the keys of the house?

'What house?' asked Miss Pym.

'Hartleys, mum, they don't put a board out, as no one is likely to pass, but it's to be sold. Miss Lefain wants to sell or let it—'

'I've just been up to see her—'

'Oh, no, mum—she's been away a year, abroad somewhere, couldn't stand the place, it's been empty since then, I just run in every day and keep things tidy—'

Loquacious and curious, the young woman had come to the fence; Miss Pym had stopped her horse.

'Miss Lefain is there now,' she said. 'She must have just come back—'

'She wasn't there this morning, mum, 'tisn't likely she'd come, either—fair scared she was, mum, fair chased away, didn't dare move her china. Can't say I've noticed anything myself, but I never stay long—and there's a smell—'

'Yes,' murmured Martha Pym faintly, 'there's a smell. What—what—chased her away?'

The young woman, even in that lonely place, lowered her voice.

'Well, as you aren't thinking of taking the place, she got an idea in her head that old Sir James—well, he couldn't bear to leave Hartleys, mum, he's buried in the garden, and she thought he was after her, chasing round them bits of china—'

'Oh!' cried Miss Pym.

'Some of it used to be his, she found a lot stuffed away, he said they were to be left in Hartleys, but Miss Lefain would have the things sold, I believe—that's years ago—'

'Yes, yes,' said Miss Pym with a sick look. 'You don't know what he was like, do you?'

'No, mum—but I've heard tell he was very stout and very old—I wonder who it was you saw up at Hartleys?'

Miss Pym took a Crown Derby plate from her bag.

'You might take that back when you go,' she whispered. 'I shan't want it, after all—'

Before the astonished young woman could answer Miss Pym had darted off across the marsh; that short hair, that earth-stained robe, the white socks. 'I generally live in the garden—'

Miss Pym drove away, breakneck speed, frantically resolving to mention to no one that she had paid a visit to Hartleys, nor lightly again to bring up the subject of ghosts.

She shook and shuddered in the damp, trying to get out of her clothes and her nostrils—that indescribable smell.

Mayday!

REDVERS BRANDLING

Captain Ian Sercombe was frightened. He rested a broad forefinger on the control column of the Boeing 747 and eased back in his seat. Glancing out of the cabin windows at the sixty metres of his giant machine's wingspan he tried to calm himself with thoughts of its size and detail . . . as high as a six storey building, over two hundred kilometres of wiring, four million parts, space for more than four hundred passengers . . .

'Decent night, Skip.'

First Officer Les Bright's voice cut in on Ian's thoughts. The two men had completed the pre take-off check and were sitting on the flight deck. Outside a huge moon hung in the hot tropical night sky which pressed down on Singapore's Changi Airport.

Les Bright was talking to the control tower when Cabin Service Director Edwina Reeves came into the flight deck area.

'Two hundred and sixty passengers and thirteen cabin crew all safely on board, Captain. Cabin secure.'

'Thanks, Edwina,' replied Ian. 'We should be off very soon.'

Minutes later the huge aircraft began to roll away from its stand at the airport. The time was 8.04pm and the journey to Perth, Australia had begun.

Within an hour all was routine on the flight deck. The Jumbo was cruising at Flight Level 370, about seven miles above sea level. Speed was 510 knots and the course was 160° magnetic as the plane, under the automatic pilot, headed south over Indonesia.

'Weather ahead looks good,' commented First Officer Bright, nodding at the weather radar screen which promised three hundred miles of smooth flying ahead.

'Hmmm,' agreed Ian.

He had been studying the weather radar with unusual intensity—just as he had all the other complex instruments in the cabin. But the fear wouldn't go away. It wasn't nervousness . . . or apprehension . . . Ian Sercombe was frightened. He could only ever remember feeling like this once before, and that had been the dreadful day of the accident . . .

Ian and his lifelong friend Mike Payne had been crewing together on a flight back from New York. Leaving the airport in Ian's car, they were accelerating on the M25 when a tyre burst. In the crash which followed Ian had been unhurt, but Mike was killed instantly. Just before the tyre went Ian had felt this unreasoning fear. Afterwards he could never quite rid himself of guilt for Mike's death. He'd been blameless perhaps—he'd checked the tyres just a couple of days previously—but how could Mike know that? Once again he thought of Mike's bluff, smiling Irish face, grinning as always and clapping those gloved hands together. Always been a joke between them that—the only pilot who never flew without wearing fine kid gloves.

Ian's thoughts were brought back to the present as First Officer Bright made a routine position report.

'Jakarta Control, Moonlight Seven over Halim at 20.44.'

Then it started.

'Unusual activity on weather radar, Captain.'

'I see it, Les.'

'Just come up—doesn't look good.'

'Could be some turbulence in that. Switch on the "Fasten Seat Belts" sign.'

The two pilots tightened their own seat belts. Behind them in the crowded cabins, passengers grumbled as they had to interrupt their evening meal to fasten their seat belts. Smiling stewardesses assured them there was no problem.

'Engine failure—Four!'

The flight engineer's terse voice cut the flight deck silence.

'Fire action Four,' responded Ian simultaneously.

Together Les Bright and Engineer Officer Mary Chalmers shut off the fuel lever to Four and pulled the fire handle. There was no fire in the engine and Ian felt an easing of his tension.

No pilot likes an engine failure, but the giant Jumbo could manage well enough on the three that were left.

'Engine failure Two.'

Mary Chalmers' voice was more urgent this time, but as she and Les Bright moved to another emergency procedure she suddenly gasped breathlessly.

'One's gone . . . and Three!'

Seven miles high with two hundred and seventy-three people on board, the Boeing was now without power. Ian knew that the huge plane could only glide—and downwards.

'Mayday, Mayday, Mayday!' First Officer Bright's voice barked into the emergency radio frequency. 'Moonlight Seven calling. Complete failure on all engines. Now descending through Flight Level 360.'

Ian's hands and mind were now working with automatic speed. He again checked the fuel and electrical systems. Emergency restarting procedures failed to have any effect. Quickly he calculated their terrible position. The plane was dropping at about two hundred feet per minute . . . which meant that in twenty-three minutes time . . .

'You two,' said Ian quietly to the First Officer and Flight Engineer. 'I'm going to need all the help I can get later on, but there could be problems back there with the passengers now— especially as we're obviously going down. Go back—help out— and get back here as soon as you can.'

Bright and Mary Chalmers climbed out of their seats, slamming the door to the flight deck behind them as they went to try and reassure the terrified passengers.

Ian was now alone on the flight deck.

'Problems,' he muttered aloud. 'Crash landing in the sea so keep the wheels up, lights are going to fail because there's no generated power from the engines, standby power from the batteries won't last long . . .'

The closing of the flight deck door interrupted Ian's monologue.

'All right back there?' he asked, as the First Officer climbed back into his seat. He was just able to make out his fellow pilot's quick nod in the rapidly dimming light on the flight deck.

'It's too risky to try and get over those mountains now,' said Ian. 'What do you think?'

'Go for the sea,' was the reply, in a strangely muffled tone.

Ian's arms were aching from holding the lurching and

buffeting aircraft, but he was surprised when the First Officer leaned over and laid a hand on his shoulder. It seemed to have both a calming and strengthening effect.

'I'll take her for a while.'

The giant plane continued to drop. At 14,000 feet the emergency oxygen masks had dropped from the roof for passengers' use. Now the rapidly dropping height was down to 13,000 feet.

'I'll save myself for the landing,' muttered Ian, watching his co-pilot in admiration. In the dim light the First Officer was a relaxed figure, almost caressing the jerking control column. His touch seemed to have calmed the aircraft too. Its descent seemed smoother, almost gentle even.

13,000.

12,000.

11,000.

'Ian.'

The captain was startled by the unexpected use of his Christian name by the First Officer.

'Volcanic dust and jet engines don't mix. I think we should make another re-light attempt on the engines now.'

Still feeling calm, even relaxed considering the terrible situation they were in, Ian began the engine restarting drill yet again.

'Switch on igniters . . . open fuel valves . . .'

As suddenly as it had failed, Engine Four sprang back into life.

'We've got a chance!' cried Ian.

'Go for the rest,' was the quiet reply.

Expertly, Ian's hands repeated the procedure. There was a lengthy pause then . . . Bingo! Number Three fired. . . then One . . . and then Two.

'We'll make it after all,' sighed Ian, once again taking a firm grip of the controls.

'Les—get on to Jakarta Control and tell them what's happening . . . Les . . .'

To his astonishment, when Ian looked to his right only the gently swaying control column came into view. The First Officer had gone. It was then that the captain heard the crash of the axe breaking through the door to the flight deck.

Engineer Chalmers was the first one through the shattered door.

'Fantastic, Skipper, fantastic—how did you do it?'

'Incredible!'

This was Les Bright's voice.

'The flight deck door jammed and we've been stuck out there for five minutes wondering how on earth you were getting on—and now this! You're a marvel, Skipper.'

Ian glanced up at the animated face of his First Officer in the brightening light of the flight deck.

'But . . . '

The rest of the words died on his lips. A feeling of inexplicable gratitude and calm swept over him. He remembered the confident, sure figure who had so recently sat in the co-pilot's seat. He now remembered too that just before the lights had reached their dimmest he had noticed that the hands holding the controls were wearing a pair of fine kid gloves.

'Get on to Jakarta,' Ian said quietly. 'Tell them we're coming in.'

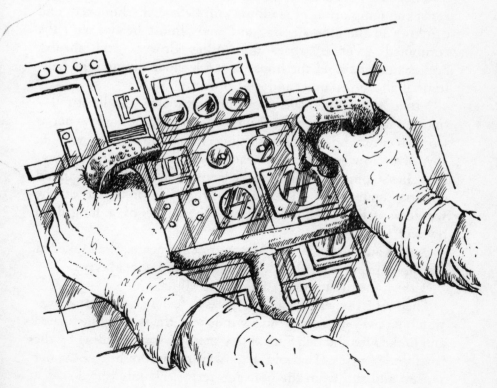

It was Rose Hall

PETRONELLA BREINBURG

The Hall was just what Nurse Minta had expected. It was a large English-looking mansion sitting on well-kept lawns, and like pictures she'd seen of the first 'old country' homes in Jamaica. Its brown brick wall stared her in the face as she walked up the gravel path, passing a shrubbery and clumps of rose bushes in full flower.

When she reached the front door Nurse Minta could not resist the temptation of standing still before she knocked, and drinking in the air—air, fragrant, and almost heady, with the combined scent of roses and other flowers and shrubs. Somewhere, beyond the house, a stable clock chimed the half-hour. It all seemed quiet and well-ordered.

Nurse Minta lifted the bright brass knocker and let it fall—once, twice. The noise made by the knocker came back to her as from an empty house, with startling clarity and force. It was the sort of hollow sound often heard in caves or echoing in great churches. Come to think of it the lawn did look a bit like an old country churchyard. The narrow rose-bed in front of the windows might well be a grave—the grave of a long-dead owner?

Bang, bang, bang! Nurse Minta knocked again. She looked at the slip of paper in her hand: The Hall, Peak Lane, Maypen.

'I *am* at the right house,' said Nurse Minta.

Then she experienced an uneasy feeling. Someone was watching her—but not from a window in the house. She turned and looked behind her. Sure as day there he was, standing by the road under a tree. That old man, waiting no doubt to hear her scream and run from the house in terror. Haunted he'd said it was when she'd asked him, a few minutes earlier, if she had reached The Hall. Haunted indeed! This well-kept, peaceful place!

Minta, occupied with her own thoughts, did not realize that the door of The Hall was opening. She gasped; the door was opening slowly, dead slow; not even a babe could have opened it as slowly as that. Now it was wide open and Minta waited for a face to appear; there was none. The door had opened seemingly on its own. A chill finger of fear crept down her spine. Beads of sweat glistened on her dark forehead.

'But, of course,' and she laughed at herself. 'That girl told me that the caretaker's wife was crippled—no doubt she has some gadget to control the door from her wheel-chair.'

So, trying to shake off any feeling of apprehension, she stepped through the doorway and into a square hall. It seemed dimly lit and cold.

'Anyone in?' Minta called, as cheerfully as she could.

'This way, please,' a pleasant voice answered.

Minta, following the sound of the voice, went down a passage, expecting to find only a pre-taped recorder that in some way was mechanically connected with the front door.

How wrong she was! She came to the entrance of a big room in which a lady sat at a small, elegant writing-table. The lady turned and said, 'I am Mrs O'Harry. Do come in and sit down.' No wheel-chair was in sight nor any mechanical contrivance. Minta wondered again how the front door had been opened.

Mrs O'Harry's face was one of the most lovely Minta had ever seen. And her whole appearance was that of a wealthy, beautiful and fastidious woman. Her grey hair hung in soft curls down to her shoulders; her dress, light gold in colour, was of pure silk, and her jewellery—a heavy gold chain, bracelet and several rings—was without doubt real.

To Nurse Minta this beautiful woman seemed strange and at the same time familiar. She thought of fancy dress, then of 'period' plays. 'Oh, of course,' said Minta to herself. 'She is like one of those portraits we saw on "art appreciation" visits to the picture gallery when I was at High School.' Looking about her, Nurse Minta thought Mrs O'Harry exactly suited the room, which she supposed must be a study or perhaps the library of the house. The walls were lined with book cases. On the writing-table two books lay open and beside them some paper on which were what looked like hand-written notes. But the paper had a yellowish tinge and the books seemed old, and almost grey—as if covered with dust.

These thoughts quickly shot through Minta's mind as she sat herself down uneasily on the edge of a chair, which had thin legs and wooden arms, and tried to think how to begin. 'It is good of you . . . You are kind to see me—I got lost—that's why I'm late. That girl—she told you I'd be coming?'

'Oh yes, she did. It does not matter that you are later than she said you would be. So, come with me now and I will show you the flat.'

Minta, obeying Mrs O'Harry's gesture, went first up the broad flight of stairs to the landing of the floor above. This floor had been altered and made into two flats. One was vacant and had been advertised as available for 'visitors to Jamaica'. Not that Minta had seen the advertisement herself. It was that girl, the friendly Jamaican girl, who had walked part of the way along Banana Grove Lane with Minta, who had told her about it.

The flat Minta could have consisted of one comfortable-looking sitting-room, a small kitchenette, and a gigantic (after what she'd been used to) well-furnished bedroom.

'You are not listening, dear,' Mrs O'Harry said, loudly. For Minta stood silent, looking about her, and had not heard Mrs O'Harry speak.

'Oh, I'm sorry. I was just thinking. It is all so nice, and just what I want . . . but . . . that girl, I thought she said only three shillings a week. Now I've seen it . . . I don't think I can . . . But did you say something?'

'I was saying that the kitchen is small—but that is right. Three shillings a week. You pay when you like. There are no conditions,' Mrs O'Harry smiled, 'and no keys either. I'll go to my room now, and leave you to have a good look round.'

Minta did have a good look round. And since she was alone, she went to try the big, comfortable-looking bed. Eh—a spring mattress! She bounced up and down. Her bed in the nurses' hostel in Kingston, where she was staying, did not bounce. Then Minta lay flat on the bed, put her arms under her head and stretched out her aching legs. Eh, what comfort! It was too good to be true. 'It must be only a dream,' murmured Minta. 'I am still in that café, trying to get up enough strength, after all my trotting round this morning, to start to climb the high hill in the hope of finding a place to stay.'

At this point Nurse Minta must have dozed off, for the next thing she heard was rain drumming against the windows. Daylight was almost gone.

Oh no! She began to panic. No umbrella, no mac, and trapped in a strange house during a typical Jamaican downpour! She must leave. Still dazed with sleep she scrambled up, stumbled in the half light into the passage, and, without realizing it, bumped against the door of the other flat. The door opened and Minta found herself face to face with a young woman, who said, 'Oh hello, you must be the new tenant.' While she spoke she looked straight at Minta with her deep-set, violet-coloured eyes.

Minta, now more or less awake, looked at her with interest and answered, 'Yes. Be moving in tomorrow.'

'You are not going out in this rain, surely?'

'I must go. Got to go back to Kingston tonight and get my things.'

'Oh, you won't need anything here. Do come in. My name is Etta.'

Minta hesitated for a moment. There was something disturbing about the young woman. Something she could not put her finger on. Her hair for example. Her hair was too golden, almost unnatural.

'But what girl has hair of natural colour today?' said Etta answering Minta's unspoken words.

Nurse Minta felt herself blush and was acutely embarrassed because her thoughts had been read. But she allowed herself to be led into the room.

A little later, sitting in a deep armchair in the girl's pleasant room, sipping a glass of ice-cold coconut water, Minta found herself telling Etta why she was in Maypen.

'I've come from the United States to see if I can discover anything about my Jamaican ancestors,' she explained.

'Oh, yes?' said Etta.

'My Dad and Mum went to Georgia and I was brought up there. I took my training in New York State, and now work in a big hospital. I'm a nurse, see? But both my grandparents were Jamaican born, and they lived right here in Maypen. That's why

I want to stay here for a bit. Even if only for two weeks, and even if I don't find out anything of the family history or discover any trace of my other ancestors, or find relations still living here.'

'It's a long way to have come,' said Etta.

'Well, but I had to save. Even in the States, young nurses don't get much. It took ages to raise the fare.'

'I used to be a nurse. All of us here. We were all nurses during our other life.'

'Your other life?'

'Yes, during my other life I was a nurse too. And Mrs O'Harry. Poor Mrs O'Harry . . . '

'What's wrong with her?'

'Nothing now. Only she was locked up here for many years. The smugglers, you see. This used to be their hideout. But you do know that, don't you? There is a secret underground passage leading right to the sandy sea-shore. Poor Mrs O'Harry was kidnapped and brought here to look after their wounded. They couldn't let her go after that, could they?'

'But that was years ago, before my grandad was born. And that was at—eh—Rose Hall. That's it. Rose Hall it was called. I was told stories about it when I was a kid.'

'But this is Rose Hall,' said Etta.

Minta laughed then. Etta was trying to frighten her. Rose Hall was haunted. Nobody dared even to talk about Rose Hall, let alone visit it.

'If this was Rose Hall,' said Minta with a big laugh, 'no one would be seen dead here. Not in the evening, anyway.'

'Some people do call us dead. They call me dead, call Mrs O'Harry dead, poor Mrs O'Harry.'

'But you're sitting right here. Of course you're not dead.'

'I'm glad *you* think so. Most people call us dead. They don't believe that we've just passed on to another world.'

Nurse Minta began laughing, stopped short. After all, mentally ill patients had to be sympathized with, Nurse Minta knew that. She understood. This poor young woman was not quite normal. She'd probably been overworking, and was at The Hall for a rest and change.

Minta got up. It was time she left, even though the storm was not quite over. Etta stood up too. Minta stretched out her hands, meaning to put them reassuringly upon the girl's shoulders:

Nurse Minta's arms remained extended, as though frozen in mid-air. Etta still stood in front of her, but Minta's hands, though apparently resting upon her shoulders, had touched . . . nothing. Minta stood, as though glued to the spot, she looked the girl up and down.

Minta gasped—she had just noticed Etta's feet. They were not even touching the floor. Minta cried out in real fright, turned, ran and almost tumbled down the stairs. Mrs O'Harry came from her room just as she reached the bottom step.

'That lodger,' gasped Minta, pointing up the stairway, 'she's . . . she's a ghost. She's a ghost,' she shouted, feeling she must make herself heard above the renewed splashing and thudding of the rain.

Mrs O'Harry spoke gently. 'Ghost? Oh dear, has Etta frightened you?'

'Yes. She's a ghost—a real ghost—a GHOST!'

'My dear, you must not upset yourself. We are not ghosts, you know, although we can't make people believe that.'

Minta stepped back. 'You . . . you? You said "we"?' She looked at Mrs O'Harry and then turned quickly to look up the stairs. The girl called Etta was still standing, or rather floating on the landing.

'Yes, dear,' came Mrs O'Harry's quiet voice. 'We. All of us here at Rose Hall.'

This was too much. 'I *am* dreaming,' said Minta aloud, 'I know, I went fast asleep on that comfortable bed.'

'You are not asleep, dear. This *is* Rose Hall.'

At these firm, though gently spoken words—Minta did indeed 'wake up'. 'Lord save us,' she muttered, 'I'm in a loony bin.' She felt that she must get out. 'Please move away from the door,' she said. And then as Mrs O'Harry's eyes seemed to take on a steely look, she spoke more roughly, and put out a hand to shove Mrs O'Harry aside. 'Out of my way—if you aren't all mad, perhaps you are a lot of crooks. I'm not staying here. Move out of my . . .'

Nurse Minta stopped. Her body became stiff with fright. She thought she had pushed Mrs O'Harry. But she had not touched anything. Her hand was just out in space—in air, vacuum—nothingness.

Nurse Minta screamed. She turned to run. And there, before her was another figure—appearing seemingly from nowhere.

Heavens, no—it could not be the same girl. Not that girl she'd walked with in the Banana Grove Lane—the girl who had given her the address of the vacant flat?

'Yes, it is.' Once more Nurse Minta's thoughts were understood, just as if she had spoken the words aloud. 'And you will be my special friend,' said the girl from the Grove floating towards Nurse Minta. 'I saw it all. It happened on the hill just in front of the house. Yes,' she repeated, almost proudly, 'I saw it all. I was in the shrubbery. It was a black car, number—number—let me think—number MV 1237. Yes, it was MV 1237, sure of it.'

'Dear God, let me wake up! Let me wake up!'

'But you are not asleep,' said the girl. 'Don't you understand? You are coming here to live with us. You and I will be great friends. I saw it all. Yes, I did.'

The Jamaican girl stopped speaking. Nurse Minta seized her chance, dashed past her and ran out of the house. Not that she thought anyone would have prevented her escape. But she somehow knew that they *could* have stopped her if they had wanted to. They had the power.

'Mad. That's it. They are all mad,' she said. Trying to reassure herself. Trying to find a rational explanation of her bewildering, terrifying experience. Then the words of the old man came back to her: 'Haunted!'

She rushed from the house into the rain, tore down the path and out on to the wet road. Two headlights bore down upon her from the top of the hill. She heard the screech of brakes. The screech came too late. Much too late. Just before that sharp impact against her body, Nurse Minta saw the number of the car: it was MV 1237.

'Not my fault,' cried a hoarse voice. 'Not my fault. She dashed out right in front of me. I tried to stop; my God, I tried to stop! She ran straight into me.'

'I saw it all.' Nurse Minta heard the voice of the girl from the Banana Grove.

She tried to speak: got to save that driver. 'It was not his fault, was not his fault.'

'No pain fortunately. Died instantly,' said the ambulance people when they came.

'Not true,' again Nurse Minta tried to speak aloud. 'I am still alive.'

No one heeded her. The ambulance crawled away. The rain
clouds had gone. All that was left on the Maypen hill was the
sweet smell of flowers. Somewhere someone was playing a
piano softly. Nurse Minta in spirit floated up and back along the
gravel path and through the door of Rose Hall.

Earthbound

ROBERTA SIMPSON BROWN

Marty got out near the church and headed down the road to the old farm he loved. He couldn't remember exactly how long he'd been away. He hadn't wanted to leave in the first place, but some things couldn't be helped. He'd had no choice.

The worst thing about leaving was the way his mother had cried. He'd been the only one left to help her work the farm, but he hadn't been able after the accident. He wouldn't be coming back now except for his mother's grieving. She hadn't actually asked him to come, but he could tell by her crying she wanted him to.

His legs felt wobbly as he walked along because he wasn't used to walking. He had to go slowly, but he didn't mind. It gave him a chance to look at the farm. Most things hadn't changed, yet something was different. It gave Marty a peculiar feeling because he couldn't figure out what it was. He felt out of place, and that disturbed him.

He knew every inch of those corn fields, and he knew he'd always be bound to that land. The wind was picking up, and he could hear the corn rustle like corn fairies whispering. He remembered how his mother used to read him that story by Carl Sandburg.

In the farmhouse down the road, Marty's mother was thinking about the old days, too. She was wishing this stormy night could be like those when the family was together. Her husband would come in from the fields, and they'd sit around the fire after supper and she'd read to Marty. After her husband died, she had still had Marty, and he had been such a comfort.

She dreaded stormy nights now that she was alone. This would be a bad one. The first clap of thunder had jarred Marty's picture right off the wall.

Back down the road, Marty tried to walk faster. He thought he'd be home by the time the storm broke, but he was very tired now. Every breath he took burned his lungs. His throat felt dry and dusty, and his shoes were covered with dirt.

He could see the shape of the old barn looming in the distance now. The combine would be inside. It was odd, but he hadn't thought of it since the accident.

He saw the farmhouse beyond and wished the black clouds wouldn't roll in so fast.

He really needed to hurry, but his body wouldn't respond like it used to.

As he passed the barn, the air felt heavy and damp and there was a strong musty odour around him. Something definitely was not right, and it worried him.

But he was almost home. He could see the light in the window, and he knew his mother would be happy to see him. He wondered if she'd think he'd changed much.

The first raindrops hit as Marty reached the yard. His energy was spent, but he forced himself to keep going. He could hardly wait to see his mother's face.

The storm began to rage around him as he climbed the steps to the porch. The wind tore at his clothes and drove him against the railing. He stumbled and grabbed the old porch swing. His weight banged it against the side of the house.

Marty's mother heard the noise, and through the window, Marty saw her move to the door to see what had happened. As the door opened, he turned towards her and gave her his biggest smile.

For a few seconds, she stood there blinking, trying to adjust to the darkness. She smelled a musty, rotten odour, and then a flash of lightning revealed the figure before her. In an instant, she saw the grinning skull and the rotting burial clothes of her dead son. She stared in disbelief as the thing from the grave reached out to her. Then Marty's mother fell forward into the skeleton's bony arms.

The Marble Hands

BERNARD CAPES

We left our bicycles by the little lych-gate and entered the old churchyard. Heriot had told me frankly that he did not want to come; but at the last moment, sentiment or curiosity prevailing with him, he had changed his mind. I knew indefinitely that there was something disagreeable to him in the place's associations, though he had always referred with affection to the relative with whom he had stayed here as a boy. Perhaps she lay under one of these greening stones.

We walked round the church, with its squat, shingled spire. It was utterly peaceful, here on the brow of the little town where the flowering fields began. The bones of the hill were the bones of the dead, and its flesh was grass. Suddenly Heriot stopped me. We were standing then to the northwest of the chancel, and a gloom of motionless trees overshadowed us.

'I wish you'd just look in there a moment,' he said, 'and come back and tell me what you see.'

He was pointing towards a little bay made by the low boundary wall, the green floor of which was hidden from our view by the thick branches and a couple of interposing tombs, huge, coffer-shaped, and shut within rails. His voice sounded odd; there was a 'plunging' look in his eyes, to use a gambler's phrase. I stared at him a moment, followed the direction of his hand; then, without a word, stooped under the heavy-brushing boughs, passed round the great tombs, and came upon a solitary grave.

It lay there quite alone in the hidden bay—a strange thing, fantastic and gruesome. There was no headstone, but a bevelled marble curb, without name or epitaph, enclosed a gravelled space from which projected two hands. They were of white marble, very faintly touched with green, and conveyed in that still, lonely spot a most curious sense of reality, as if actually thrust up, deathly and alluring, from the grave beneath. The impression grew upon me as I looked, until I could have thought they moved stealthily, consciously, turning in the soil as if to greet me. It was absurd, but—I turned and went rather hastily back to Heriot.

'All right. I see they are there still,' he said; and that was all. Without another word we left the place and, remounting, continued our way.

Miles from the spot, lying on a sunny downside, with the sheep about us in hundreds cropping the hot grass, he told me the story:

'She and her husband were living in the town at the time of my first visit there, when I was a child of seven. They were known to Aunt Caddie, who disliked the woman. I did not dislike her at all, because, when we met, she made a favourite of me. She was a little pretty thing, frivolous and shallow; but truly, I know now, with an abominable side to her.

'She was inordinately vain of her hands; and indeed they were the loveliest things, softer and shapelier than a child's. She used to have them photographed, in fifty different positions; and once they were exquisitely done in marble by a sculptor, a friend of hers. Yes, those were the ones you saw. But they were cruel little hands, for all their beauty. There was something wicked and unclean about the way in which she regarded them.

'She died while I was there, and she was commemorated by her own explicit desire after the fashion you saw. The marble hands were to be her sole epitaph, more eloquent than letters. They should preserve her name and the tradition of her most exquisite feature to remoter ages than any crumbling inscription could reach. And so it was done.

'That fancy was not popular with the parishioners, but it gave me no childish qualms. The hands were really beautifully modelled on the originals, and the originals had often caressed me. I was never afraid to go and look at them, sprouting like white celery from the ground.

'I left, and two years later was visiting Aunt Caddie a second time. In the course of conversation I learned that the husband of

the woman had married again—a lady belonging to the place—
and that the hands, only quite recently, had been removed. The
new wife had objected to them—for some reason perhaps not
difficult to understand—and they had been uprooted by the
husband's order.

'I think I was a little sorry—the hands had always seemed
somehow personal to me—and, on the first occasion that
offered, I slipped away by myself to see how the grave looked
without them. It was a close, lowering day, I remember, and the
churchyard was very still. Directly, stooping under the
branches, I saw the spot. I understood that Aunt Caddie had
spoken prematurely. The hands had not been removed so far,
but were extended in their old place and attitude, looking as if
held out to welcome me. I was glad; and I ran and knelt, and put
my own hands down to touch them. They were soft and cold
like dead meat, and they closed caressingly about mine, as if
inviting me to pull—to pull.

'I don't know what happened afterwards. Perhaps I had been
sickening all the time for the fever which overtook me. There
was a period of horror, and blankness—of crawling, worm-
threaded immurements and heaving bones—and then at last the
blessed daylight.'

Heriot stopped, and sat plucking at the crisp pasture.

'I never learned,' he said suddenly, 'what other experiences
synchronized with mine. But the place somehow got an
uncanny reputation, and the marble hands were put back.
Imagination, to be sure, can play strange tricks with one.'

Fat Andy

STEPHEN DUNSTONE

Andrew and his mother were walking through St Leonard's churchyard, on their way back from the shops. It was a short cut they had been taking for the last forty years or so, and it was pleasanter than going by the road: there was the well-kept grass, the beautifully trimmed hedge, the daffodils in spring, the swallows nesting over the south transept window. . .

'That window's going to be filthy before long,' said Millie.

But Andrew's mind was not on swallows—he was grappling with a difficult thought. For the church was to be made redundant, and the important issue for Andrew—the thing Andrew was trying to puzzle out—was this: when would God's spirit actually leave? Would it be when the Bible and prayer books were removed, or would He wait until the roof fell in? Andrew had special reasons for wanting to know.

'I saw a damp patch in the vestry this morning,' Millie was saying. 'It's a leak in the plumbing, I expect. Wasn't the roof, we haven't had rain. Might be the mains, under the floor.'

'Mains is off,' said Andrew.

'Doesn't mean you can't get water,' said Millie. 'You don't get water when the mains is *on* sometimes, so you could easily get it when it's off. You should look.'

'God's going,' said Andrew. 'Bad'll come.'

'Don't talk daft. God doesn't want dafties.'

He didn't answer.

'Oh, now would you look at that!' exclaimed Millie, as they passed a well-tended grave. 'Is that the cats been playing around? They've scattered all the flowers you put out. Would you believe it?'

But Andrew was panting away down the path, the shopping bag swinging against his legs.

He shut the church door behind him—rather loudly in his haste—and stood on the worn stone step, getting his breath back. He looked around the old walls, seeking reassurance. Yes: God was still here.

Safe now, he could let the pictures come into his head. They'd been coming a lot recently.

In the pictures a wave crashes on a rocky beach; two children are playing among the pools: a girl and a boy. No one else is in sight. The boy does not seem to be enjoying himself.

Their voices—her voice—is carried on the breeze: 'Can't catch me! Can't catch me!' Then, with a sudden burst of inspired rhyming: ''Cos you're much too FAT, got a face like a big cow PAT!'

But he has almost caught her. He lunges—a fat boy's lunge—she squeals and runs off across the rocks. 'FatAndyfatAndy fatAndyfatAndy . . . can't catch me . . .'

It seems he doesn't want to play any more; he turns to go back the way they came.

Instantly: 'Where are you going?' calls the girl. 'Come back! Spoil sport!'

But he plods resolutely along the shingle. The girl runs to catch him up.

'I didn't mean it,' she says. 'Please play.'

He ignores her.

'Honestly, you're really nice, I really think so.'

He falters.

She dangles bait. 'I want to show you something.' He stops, looks at her. 'It's a secret, it's a really special secret. Do you want to know?'

His look tells her he does, very much.

'Then you've got to catch me. I won't run too fast.' And she's off again; off down the beach and onto a rock, where—coquettishly, like some shore-bird in a mating display—she calls: 'Andy . . . Andy . . . Andy . . . Andy . . .'

He makes up his mind—he cannot live without the secret—and he trots after her. Instantly, she's away into the distance, calling into the wind: 'FatAndyfatAndyfatAndyfatAndy . . .'

The pictures faded, and Andrew blinked in the sunlight that shafted through the south transept window, undimmed by swallows' droppings.

He whispered to God: 'It wasn't cats. I've kept her sleeping, till now, but she knows, see. She knows you're going.'

The next day, Andrew was in the church again, with his mother, collecting hassocks and piling them up. Millie was looking out for the ones she had embroidered; Andrew was wondering if a bargain could be struck with God. If they left a hassock, and Andrew came and knelt on it, and if they left a prayer book and a hymn book somewhere, couldn't He be persuaded to stay? It wouldn't be like before, but He could pretend. Couldn't He?

'Aren't you going to go and look at that damp?' said Millie. 'It might be doing something in the crypt.'

'Not my job any more,' said Andrew petulantly.

'Someone'll have to look: you don't want a flood.' They carried on piling. 'I hope it's fine for the vicar's little speech tomorrow,' said Millie, after a while.

'Is it after that, God goes?' said Andrew. 'Is it what the vicar says?'

'It's nothing to do with the vicar, it's the bishop signing a paper, I expect.'

'I hope he never signs it,' said Andrew.

'You're soft in the head,' said his mother.

Outside, somewhere, a dog barked. Andrew dropped the hassock he was holding and looked up, ready for flight. 'What was that?'

'A dog, silly.'

'Wild dog?'

'Quite soft in the head,' said Millie. But he didn't hear her.

The girl and boy are walking side by side, close to the water, talking.

'Why are you fat?' asks the girl.

'Don't know,' says the boy, simply.

'I expect they could operate,' says the girl. 'My father says they can do anything in hospital. I expect they could slice bits off.'

'Don't want to go to hospital.'

'They'll have machines they can bring and do it at home for you.'

'Don't want machines.'

'I'd hate to be fat,' she says. 'I'm glad I'm beautiful. I won our class beauty competition. It's my bone structure.'

'Where are we going?' asks the boy. 'What's the secret you're going to show me?'

'Do you want to hold my hand?'

He thinks about it. 'All right.'

'Well you can't!' And she runs off again. ''Cos you're too fat! FatAndyfatAndyfatAndyfatAndy . . .'

At the appointed time, a handful of ancient faithfuls gathered in the churchyard to hear the vicar's little speech and see Andrew get his present. All those years of keeping the churchyard looking so nice: it was only proper he should get his thanks. The sun shone benignly down.

'His whole life,' the vicar was saying, 'has been dedicated to the service of simple beauty: witness his devotion to the remembrance of young Joanne, who was taken so untimely from her family all those many, many years ago. It is Andrew in his unquestioning faith and selflessness who has acted as a model to us all, tending the memory of a blessed innocent with the flowers of God's garden. I would love to say "Long may it all continue thus", but alas . . .'

And on he went, talking about dwindling congregations, straitened times, and the recent problem of subsidence. But Andrew wasn't listening. He was staring with widening eyes at a certain headstone. It was at a tilt, surely it was at a tilt. It hadn't been yesterday. His breath began to come fast and shallow. He turned, blundered past his mother, past the vicar and ancient faithfuls.

'Andrew!' bellowed his mother. 'Come back here, this is for you!' But he was gone, into his refuge.

Once inside, he leant his head against the wall, waiting for his pounding heart to ease. He laid his cheek against the cool plaster, touched its solidity with his hands. He stood there without moving.

The girl must have waited for him to catch up, because they're walking together again.

'It must be very boring being called Andrew,' she says.

But he doesn't reply. So much of what she says confuses him. They walk on. Looking straight ahead of her, the girl says, 'I think you're in love with me.' His heart beats faster, but again he says nothing, and she rounds on him, almost ferociously: 'Are you?'

'Where are we?' says the boy. 'I don't know where we are.'

'You're stupid. I don't think I'm going to show you my secret any more,' she says, carelessly.

'Please.'

'I'm going to push you in the sea and watch you float away to America.'

He looks anxiously at her. 'Don't push me in.'

'I will. After three. One—'

'Don't!' He starts to run off. 'I'm going back, I don't want to see your secret, you're horrid.'

'No, I'm not, please, Andrew,' she calls. 'I won't be horrid again, cross my heart and hope to die, I promise I won't push you in.'

'I'll push you in.' But he has stopped running.

They stand on the rocks, facing each other.

'Shall we play kiss chase?' says Joanne.

Andrew sat at the foot of the stairs, staring unseeing at a spot on the hall carpet. The vicar'd said the church would have to be padlocked from tomorrow, and Andrew'd got upset, so his mother had had words with him: told him to pull himself together. But he *was* together, he didn't need pulling; she didn't understand; she didn't *know*. He didn't like it when his mother scolded him.

He sat up, aware suddenly that he could see her reflection in the hall mirror, standing on the stairs above him. He'd thought she was outside.

'Mother?' he said, to the reflection.

A voice that was not her voice spoke softly from behind him: 'Andrew . . .'

He leapt up in panic, turned to face the voice, but there was no one there. He rushed along the hall, through the kitchen and out of the back door.

'Mother, Mother, Mother,' he cried, in the garden.

'Yes, dear, I'm here, what's happened?'

'She . . . she . . . she . . . she . . . she . . . she . . .'

'Shh, there . . .'

'She . . . she . . .'

Millie put her arm around his quivering shape. 'Who, dear?'

'She was in there, she was you.'

'Who was, dear?'

'Joanne, she was pretending to be you. Is *this* you? You're not her still?'

'No, dear, I'm always me and no one else, especially when I'm picking beans.'

He buried his face in her shoulder. 'She's the Devil,' he murmured.

'Now, now, Andrew, she's your little friend.'

He looked at his mother earnestly. 'Never let her be you. She mustn't get in.'

The boy and girl have left the beach: she has led him into the most beautiful place in the world, where the sound of water dripping in solitary drops into a rock pool echoes off every facet of the vaulted chamber, and where the sunlight reflects in ripples across the rock ceiling.

She talks in a hushed voice. 'Isn't it lovely?'

'It's lovely,' he agrees. He likes her again, now.

'I discovered it, and no one else knows about it.'

He feels like her only friend. 'It's like heaven,' says the boy.

But she has to spoil everything. 'I don't believe in heaven,' she says.

The boy is upset. 'You have to!'

'I don't believe in hell either.'

'But you have to!'

'Why?'

''Cos . . . 'cos . . . when you die.'

'Huh.' She's contemptuous. 'When I die I shall rot away, like everyone else.'

It disturbs him, what she says. ''Tisn't like that!'

'And when I'm buried, wild dogs will come and dig up my grave and my ghost will come and haunt you!'

'No! No!' cries the boy.

But she climbs nimbly over the rocks towards the obscurity in the recesses of the cave. 'It's even better further in, it gets spooky.'

'Wait! Don't leave me alone,' he calls, fearfully. 'Joanne . . .! Joanne . . .?'

His voice echoes unanswered.

Andrew looked up from where he sat on the sanctuary step. The vicar was standing quietly at the back.

'Vicar?' said Andrew. The vicar walked slowly up the aisle and stopped in front of him.

'Andrew, old chap,' he said.

'Don't let the bishop sign the paper,' said Andrew.

But it was not in the vicar's hands. He sat beside Andrew, and talked gently with him for about half an hour. Afterwards they both walked across to Andrew's house, where Millie had the kettle on.

The vicar and Millie chatted and drank their tea, while Andrew munched custard creams and pondered the vicar's earlier words. He didn't exactly understand what a psychological anxiety transference was, but the vicar, who knew about such things, had assured him that the Devil was not involved. Without wishing to get too technical (the vicar had said) it basically meant that Andrew had a fertile imagination, and though, goodness me, there was nothing wrong with imagination, no, it simply meant that these recent . . . worries had been caused by too much *thinking*. However, there were two things Andrew had to be completely clear about: firstly, there were no such things as ghosts (except the Holy Ghost, and that was something different), and secondly, even if there were, they couldn't possibly appear as another person. Categorically not.

Andrew took another custard cream. He felt curiously elated. It wasn't real, none of it was real. If he told it all to go away, it would. It was all in his head. All he had to do was say to himself, 'It's my imagination' (this was the vicar's suggestion), and it would disappear, pouff!

That evening Millie had a WI meeting. It was raining when she returned. She stamped her wet feet on the mat and shook her raincoat.

'Wet,' said Andrew, who had come to greet her in the hall.

'I don't know about locking the church,' she said, 'they should padlock the churchyard. Wretched animals.'

'What animals?' said Andrew.

'Scuffling, digging, scratching up the earth, never did like dogs. Shooed them off though. Mind out of the way, I'll have to change these shoes.'

Andrew froze. Wild dogs . . . When Millie had stomped upstairs, he tiptoed to the sitting-room window, lifted the corner of the curtain and looked churchwards. What a fool he'd been. Oh what a fool! He'd let himself be taken off guard, and now she was free, she was out. He could feel the numbness of terror taking hold inside his head.

He looked round the room. Here was obviously no longer safe: she would be coming for him any moment. The question was: could he make it to the church if he made a dash for it now, or would she be waiting for him to do just that, waiting in the darkness in the bushes beside the path? He took a deep breath and ran out of the house.

It was raining hard, but he scarcely noticed it. He opened the garden gate and lumbered out on to the church path. The bushes seemed full of menace, but he stared ahead of him as he ran, fixing his eyes on the south porch.

He reached the church, flung open the door, slammed it behind him and almost fell down the step. He knew the electricity was off, but he didn't need lights: not in here. He made his way forward to the front pew, and sat, relieved. He'd made it.

Now that the immediate panic was over, he made his plan. What he was going to do was this: he was going to spend all night here, and when they came with the padlock he'd take it away from them. Then he'd live here for ever. For ever.

The cave goes back a long way, narrowing, twisting and opening again into gloomy, chill places. The boy has followed the girl's voice into darkness, and now he has lost her.

'Joanne . . .?' he asks. 'Joanne . . .?'

Her voice comes from where he did not expect it, some distance away. 'Here I am.'

'Where? I can't see you.'

'You should eat more carrots, and not so many sweets.' The voice seems to move about in the darkness.

'I'm going,' says the boy. 'I hate you.'

'I thought you wanted to kiss me,' says the girl's voice, from somewhere else. He doesn't answer. 'Don't you?'

'You're ugly,' he says.

'Not as ugly as you. You ought to be put down. We had a puppy once that was really ugly, and I drowned it. It was easy. I could drown you too, easy-peasy. Just leave you in here and wait for the tide to come in.'

'No you couldn't, you're not going to touch me. I'm going.'

'You don't know the way out, fatty.'

'Do!' he says. He hates her. He hates her.

Instantly there is the sound of scrambling, and of falling stones, muddled with its own echo, then silence.

'Joanne . . .?' says the boy, afraid.

There is no sound, then, suddenly, close to his ear: 'Booh!'

He lashes out, seizes what his hands find.

'Ow, stop it, get off, stop it, Andrew!' She wrenches herself free and runs crunching across pebbles. 'FatAndyfatAndyfatAndy fatAndyfatAndy . . .'

He grabs whatever is at his feet—small stones, shingle—and flings it in fistfuls at the sound of the voice.

'Don't! Stop it! Ow, Andrew! Don't!'

He picks up bigger pebbles, flings them, panting, a constant barrage, not stopping, throwing indiscriminately, heavy stones, crying because he only wanted to be her friend.

'Don't, Andrew, don't!' She's almost screaming. 'Andrew, don't, don't, don't, don't, don't . . .! Andrew! Andrew! Andr—'

Her cries stop abruptly, but he does not seem to notice. He continues flinging stones, panting heavily. Then he turns and stumbles across the pebbles. He yells back to her: 'And I do know the way out. I do. I do!' And he is gone: returned to the sunlight and the sound of the sea breaking on the shore.

Andrew heard the church door open far behind him. He turned round. Torchlight bobbed.

'Andrew?' said his mother's voice.

'Mother,' said Andrew.

'What are you doing here?'

He could tell she was approaching, from the footsteps and the way the light moved. The footsteps came right up to him.

He thought. 'I came to look at the leak,' he said.

'At this time of night?' She turned and walked away towards the crypt stairway door.

Reluctantly, he got up and walked after the bobbing light.

She opened a door; listened. 'It's doing well.'

Andrew caught up. 'Sounds bad. Shine the torch.' But the stone stairs twisted out of sight.

'Down you go,' said Millie.

'We should get the vicar. We can't do anything.'

'Nonsense. On you go. You'll need your hands—I'll hold the torch.' She lit the first few steps for him.

'I don't like it, it's steep,' said Andrew. He reached the bend. 'Are you coming? You've got to come, Mother.' The light went out. 'Mother! What have you done? Turn it on!' There was no answer. 'I can't see!'

From the blackness at the top of the stone stairs came the soft voice that was not his mother's. 'You should eat more carrots and not so many sweets. I told you before.'

The chill went through him. 'Mother? Where are you?'

'She's not here,' said the soft voice. 'She's never been here.'

'She came in just now.'

'No,' said the voice. 'That was me.'

'You can't do that,' said Andrew. ''Tisn't possible. Vicar said.'

'The vicar!'

He tried the magic words. 'It was 'magination.'

'Shows how much he knows.'

'And it wasn't you last time, in the house. It was 'magination.'

'Oh yes, *that* was your imagination. But it was such a good idea I couldn't resist it. Do you mind?' The voice seemed to be coming closer, down the stairs, coming closer through the darkness, floating invisibly towards him. He backed down a step.

'You're not real,' he cried. 'You're not real! God, she's not here, tell her to go away! God?'

'He's not here either. He went this afternoon.' The voice was close now. 'Did you think He was staying till tomorrow?'

'You're not here, you're not here!'

The voice stopped an inch from his face. 'Don't you want to kiss me?'

Andrew leaned back from the touch of the long dead lips, slipped down a step. 'Don't come near me, leave me alone!'

But the voice was so close he could almost feel its breath. 'Kiss me . . .'

'No!' he cried. He lurched away from the terrible embrace, tripped, stumbled, lost balance, flung his hands out to save himself but grasped only air, and fell . . .

There were three inches of water in the crypt, but Andrew's head made a sharp crack as it hit the stone. He lay awkwardly, twisted, his legs half on the bottom steps, his face down.

The water lapped his cheek. He didn't move.

Her feet nice and dry, Millie came downstairs. The front door was open again. 'Andrew?' she called, peering out into the rain from the doormat. 'Are you out there? What are you playing at?' No answer. She looked in the sitting-room, in the kitchen. Back at the front door, she called again: 'Andrew? I'm making some cocoa, I'm putting it on now. You know you don't like it with skin on.' That'll bring him if he's coming, she thought.

She pushed the door to and went to get the milk out of the fridge.

Somewhere else, the tide is coming in. A wave races foaming up the beach, across the rocks, and into a cave at the base of the cliff. A few moments later the water surges back through the pebbles to rejoin the sea and gather itself for the next wave. An hour later there is no cave to be seen.

When morning comes, the sea has retreated and the sun shines down once more. The rocks dry in the warmth of the new day.

Bang, Bang
–Who's Dead?

JANE GARDAM

There is an old house in Kent not far from the sea where a little ghost girl plays in the garden. She wears the same clothes winter and summer—long black stockings, a white dress with a pinafore, and her hair flying about without a hat, but she never seems either hot or cold. They say she was a child of the house who was run over at the drive gates, for the road outside is on an upward bend as you come to the gates of The Elms—that's the name of the house, The Elms—and very dangerous. But there were no motor cars when children wore clothes like that and so the story must be rubbish.

No grown person has ever seen the child. Only other children see her. For over fifty years, when children have visited this garden and gone off to play in it, down the avenue of trees, into the walled rose-garden, or down deep under the high dark caves of the polished shrubs where queer things scutter and scrattle about on quick legs and eyes look out at you from round corners, and pheasants send up great alarm calls like rattles, and whirr off out of the wet hard bracken right under your nose, 'Where've you been?' they get asked when they get back to the house.

'Playing with that girl in the garden.'

'What girl? There's no girl here. This house has no children in it.'

'Yes it has. There's a girl in the garden. She can't half run.'

When last year The Elms came up for sale, two parents—the parents of a girl called Fran—looked at each other with a great longing gaze. The Elms.

'We could never afford it.'

'I don't know. It's in poor condition. We might. They daren't ask much for such an overgrown place.'

'All that garden. We'd never be able to manage it. And the house is so far from anywhere.'

'It's mostly woodland. It looks after itself.'

'Don't you believe it. Those elms would all have to come down for a start. They're diseased. There's masses of replanting and clearing to do. And think of the upkeep of that long drive.'

'It's a beautiful house. And not really a huge one.'

'And would you *want* to live in a house with—'

They both looked at Fran who had never heard of the house. 'With what?' she asked.

'Is it haunted?' she asked. She knew things before you ever said them, did Fran. Almost before you thought them.

'Of course not,' said her father.

'Yes,' said her mother.

Fran gave a squealing shudder.

'Now you've done it,' said her father. 'No point now in even going to look at it.'

'How is it haunted?' asked Fran.

'It's only the garden,' said her mother. 'And very *nicely* haunted. By a girl about your age in black stockings and a pinafore.'

'What's a pinafore?'

'Apron.'

'*Apron*. How cruddy.'

'She's from the olden days.'

'Fuddy-duddy-cruddy,' said Fran, preening herself about in her T-shirt and jeans.

After a while though she noticed that her parents were still rattling on about The Elms. There would be spurts of talk and then long silences. They would stand for ages moving things pointlessly about on the kitchen table, drying up the same plate three times. Gazing out of the windows. In the middle of Fran's telling them something about her life at school they would say suddenly, 'Rats. I expect it's overrun with rats.'

Or, 'What about the roof?'

Or, 'I expect some millionaire will buy it for a Country Club. Oh, it's far beyond us, you know.'

'When are we going to look at it?' asked Fran after several

days of this, and both parents turned to her with faraway eyes.

'I want to see this girl in the garden,' said Fran because it was a bright sunny morning and the radio was playing loud and children not of the olden days were in the street outside, hurling themselves about on bikes and wearing jeans and T-shirts like her own and shouting out 'Bang, bang, you're dead.'

'Well, I suppose we could just telephone,' said her mother. 'Make an appointment.'

Then electricity went flying about the kitchen and her father began to sing.

They stopped the car for a moment inside the propped-back iron gates where there stood a rickety table with a box on it labelled 'Entrance Fee. One pound.'

'We don't pay an entrance fee,' said Fran's father. 'We're here on business.'

'When I came here as a child,' said Fran's mother, 'we always threw some money in.'

'Did you often come?'

'Oh, once or twice. Well yes. Quite often. Whenever we had visitors we always brought them to The Elms. We used to tell them about—'

'Oh yes. Ha-ha. The ghost.'

'Well, it was just something to do with people. On a visit. I'd not be surprised if the people in the house made up the ghost just to get people to come.'

The car ground along the silent drive. The drive curved round and round. Along and along. A young deer leapt from one side of it to the other in the green shadow, its eyes like lighted grapes. Water in a pool in front of the house came into view.

The house held light from the water. It was a long, low, creamy-coloured house covered with trellis and on the trellis pale wisteria, pale clematis, large papery early roses. A huge man was staring from the ground-floor window.

'Is that the ghost?' asked Fran.

Her father sagely, solemnly parked the car. The air in the garden for a moment seemed to stir, the colours to fade. Fran's mother looked up at the gentle old house.

'Oh—look,' she said, 'it's a portrait. Of a man. He seems to be looking out. It's just a painting, for goodness sake.'

But the face of the long-dead eighteenth-century man eyed the terrace, the semi-circular flight of steps, the family of three looking up at him beside their motor car.

'It's just a painting.'

'Do we ring the bell? At the front door?'

The half-glassed front door above the staircase of stone seemed the door of another shadowy world.

'I don't want to go in,' said Fran. 'I'll stay here.'

'Look, if we're going to buy this house,' said her father, 'you must come and look at it.'

'I want to go in the garden,' said Fran. 'Anyone can see the house is going to be all right.'

All three surveyed the pretty house. Along the top floor of it were heavily-barred windows.

'They barred the windows long ago,' said Fran's mother, 'to stop the children falling out. The children lived upstairs. Every evening they were allowed to come down and see their parents for half an hour and then they went back up there to bed. It was the custom for children.'

'Did the ghost girl do that?'

'Don't be ridiculous,' said Fran's father.

'But did she?'

'What ghost girl?' said Fran's father. 'Shut up and come and let's look at the house.'

A man and a woman were standing at the end of the hall as the family rang the bell. They were there waiting, looking rather vague and thin. Fran could feel a sort of sadness and anxiety through the glass of the great door, the woman with her gaunt old face just standing; the man blinking.

In the beautiful stone hall at the foot of the stairs the owners and the parents and Fran confronted each other. Then the four grown people advanced with their hands outstretched, like some old dance.

'The house has always been in my family,' said the woman. 'For two hundred years.'

'Can I go out?' asked Fran.

'For over fifty years it was in the possession of three sisters. My three great-aunts.'

'Mum—can I? I'll stay by the car.'

'They never married. They adored the house. They scarcely ever left it or had people to stay. There were never any children in this house.'

'Mum—'

'*Do*,' said the woman to Fran. 'Do go and look around the garden. Perfectly safe. Far from the road.'

The four adults walked away down the stone passage. A door to the dining-room was opened. 'This,' said the woman, 'is said to be the most beautiful dining-room in Kent.'

'What was that?' asked Fran's mother. 'Where is Fran?'

But Fran seemed happy. All four watched her in her white T-shirt running across the grass. They watched her through the dining-room window all decorated round with frills and garlands of wisteria. 'What a sweet girl,' said the woman. The man cleared his throat and went wandering away.

'I think it's because there have never been any children in this house that it's in such beautiful condition,' said the woman. 'Nobody has ever been unkind to it.'

'I wouldn't say,' said Fran's mother, 'that children were—'

'Oh, but you can tell a house where children have taken charge. Now your dear little girl would never—'

The parents were taken into a room that smelled of rose-petals. A cherry-wood fire was burning although the day was very hot. Most of the fire was soft white ash. Somebody had been doing some needlework. Dogs slept quietly on a rug. 'Oh, Fran would love—' said Fran's mother looking out of the window again. But Fran was not to be seen.

'Big family?' asked the old man suddenly.

'No. Just—Just one daughter, Fran.'

'Big house for just one child.'

'But you said there had never been children in this house.'

'Oh—wouldn't say never. Wouldn't say never.'

Fran had wandered away towards the garden but then had come in again to the stone hall, where she stopped to look at herself in a long dim glass. There was a blue jar with a lid on a low table, and she lifted the lid and saw a heap of dried rose-petals. The lid dropped back rather hard and wobbled on the jar as if to fall off. 'Children are unkind to houses'; she heard the floating voice of the woman shepherding her parents from one room to another.

Fran pulled an unkind face at the jar. She turned a corner of the hall and saw the staircase sweeping upwards and round a corner too. On the landing someone seemed to be standing, but then as she looked seemed not to be there after all.

'Oh yes,' she heard the woman's voice, 'oh yes, I suppose so. Lovely for children. The old nurseries should be very adequate. We never go up there.'

'If there are nurseries,' said Fran's father, 'there must once have been children.'

'I suppose so. Once. It's not a thing we ever think about.'

'But if it has always been in your family it must have been inherited by children?'

'Oh, cousins. Generally cousins inherited. Quite strange how children have not been actually born here.' Fran, who was sitting outside on the steps now in front of the open door, heard the little group of grown-ups clatter off along the stone pavement to the kitchens and thought, 'Why are they going on about children so?'

She thought, 'When they come back I'll go with them. I'll ask to see that painted man down the passage. I'd rather be with Mum to see him close.'

Silence had fallen. The house behind her was still, the garden in front of her stiller. It was the moment in an English early-summer afternoon when there is a pause for sleep. Even the birds stopped singing. Tired by their almost non-stop territorial squawks and cheeps and trills since dawn, they declare a truce and sit still upon branches, stand with heads cocked listening, scamper now and then in the bushes across dead leaves.

When Fran listened very hard she thought she could just hear the swish of the road, or perhaps the sea. The smell of the early roses was very strong. Somewhere upstairs a window was opened and a light voice came and went as people moved from room to room. 'Must have gone up the back stairs,' Fran thought and leaned her head against the fluted column of the portico. It was strange. She felt she knew what the house looked like upstairs. Had she been upstairs yet or was she still thinking of going? Going. Going to sleep. Silly.

She jumped up and said, 'You can't catch me. Bang, bang—you're dead.'

She didn't know what she meant by it so she said it again out loud. 'Bang, bang. You're dead.'

She looked at the garden, all the way round from her left to her right. Nothing stirred. Not from the point where a high wall stood with a flint arch in it, not on the circular terrace with the round pond, not in the circle of green with the round gap in it where the courtyard opened to the long drive, and where their car was standing. The car made her feel safe.

Slowly round went her look, right across to where the stone urns on the right showed a mossy path behind them. Along the path, out of the shadow of the house, sun was blazing and you could see bright flowers.

Fran walked to the other side of the round pond and looked up at the house from the courtyard and saw the portrait again looking out at her. It must be hanging in a very narrow passage, she thought, to be so near to the glass. The man was in some sort of uniform. You could see gold on his shoulders and lace on his cuffs. You could see long curls falling over his shoulders. Fancy soldiers with long curls hanging over their uniform. Think of the dandruff.

'Olden days,' said Fran, 'bang, bang, you're dead,' and she set off at a run between the stone urns and in to the flower garden. 'I'll run right round the house,' she thought. 'I'll run like mad. Then I'll say I've been all round the garden all by myself, and not seen the ghost.'

She ran like the wind all round, leaping the flowerbeds, tearing along a showering rose-border, here and there, up and down, flying through another door in a stone wall among greenhouses and sheds and old stables, out again past a rose-red dove-house with the doves like fat pearls set in some of the little holes, and others stepping about the grass. Non-stop, non-stop she ran, across the lawn, right turn through a yew hedge, through the flint arch at last and back to the courtyard. 'Oh yes,' she would say to her friends on their bikes. 'I did. I've been there. I've been all round the garden by myself and I didn't see a living soul.'

'A *living* soul.'

'I didn't see any ghost. Never even thought of one.'

'You're brave, Fran. I'd never be brave like that. Are your parents going to buy the house?'

'Don't suppose so. It's very boring. They've never had any children in it. Like an old-folks home. Not even haunted.'

Picking a draggle of purple wisteria off the courtyard wall—

and pulling rather a big trail of it down as she did so—Fran began to do the next brave thing: to *walk* round the house. Slowly. She pulled a few petals of the wisteria and gave a carefree sort of wave at the portrait in the window. In front of it, looking out of the window, stood a little girl.

Then she was gone.

For less than a flick of a second Fran went cold round the back of the neck. Then hot.

Then she realized she must be going loopy. The girl hadn't been in a pinafore and frilly dress and long loose hair. She'd been in a white T-shirt like Fran's own. She had been Fran's own reflection for a moment in the glass of the portrait.

'Stupid. Loopy,' said Fran, picking off petals and scattering them down the mossy path, then along the rosy flagstones of the rose-garden. Her heart was beating very hard. It was almost pleasant, the fright and then the relief coming so close together.

'Well, I thought I saw the ghost but it was only myself reflected in a window,' she'd say to the friends in the road at home.

'Oh Fran, you are brave.'

'How d'you know it was you? Did you see its face? Everyone wears T-shirts.'

'Oh, I expect it was me all right. They said there'd never been any children in the house.'

'What a cruddy house. I'll bet it's not true. I'll bet there's a girl they're keeping in there somewhere. Behind those bars. I bet she's being imprisoned. I bet they're kidnappers.'

'They wouldn't be showing people over the house and trying to sell it if they were kidnappers. Not while the kidnapping was actually going on, anyway. No, you can tell—' Fran was explaining away, pulling off the petals. 'There wasn't anyone there but me.'

She looked up at the windows in the stable-block she was passing. They were partly covered with creeper, but one of them stood open and a girl in a T-shirt was sitting in it, watching Fran.

This time she didn't vanish. Her shiny short hair and white shirt shone out clear. Across her humped-up knees lay a comic. She was very much the present day.

'It's you again,' she said.

She was so ordinary that Fran's heart did not begin to thump

at all. She thought, 'It must be the gardener's daughter. They must live over the stables and she's just been in the house. I'll bet she wasn't meant to. That's why she ducked away.'

'I saw you in the house,' Fran said. 'I thought you were a reflection of me.'

'Reflection?'

'In the picture.'

The girl looked disdainful. 'When you've been in the house as long as I have,' she said, 'let's hope you'll know a bit more. Oil paintings don't give off reflections. They're not covered in glass.'

'We won't be keeping the oil paintings,' said Fran grandly. 'I'm not interested in things like that.'

'I wasn't at first,' said the girl. 'D'you want to come up? You can climb over the creeper if you like. It's cool up here.'

'No thanks. We'll have to go soon. They'll wonder where I am when they see I'm not waiting by the car.'

'Car?' said the girl. 'Did you come in a car?'

'Of course we came in a car.' She felt furious suddenly. The girl was looking at her oddly, maybe as if she wasn't rich enough to have a car. Just because she lived at The Elms. And she was only the gardener's daughter anyway. Who did she think she was?

'Of course we're going home by car.'

'Well take care on the turn-out to the road then. It's a dangerous curve. It's much too hot to go driving today.'

'I'm not hot,' said Fran.

'You ought to be,' said the girl in the T-shirt, 'with all that hair and those awful black stockings.'

Carlotta

ADÈLE GERAS

Diary of Edward Stonely, submitted in evidence at the Coroner's Inquest, 15 May 1993

My doctor (I refuse to call him 'my shrink', although that's what he is. It seems like an admission of madness.) has said that the dreams might stop altogether if I write everything down. It would be, he suggested, a kind of purging. It would clear my system of what he calls 'unresolved guilts.'

I felt a fool consulting him in the first place, but my dreams were becoming so dreadful that I was deliberately keeping myself awake for as long as I could every night. This meant that I was irritable and moody at school, tearing strips off both pupils and colleagues for no good reason. I was also horrible at home, to my wife Annie, whom I love more than anyone except my little daughter, Beth. I had not yet reached the point of taking my moods out on a four year old, but I can't have been the pleasantest dad in the world. The worst of it was, however hard I tried to stay awake, my eyelids closed in the end. They always did, every night, and every night, there she was: Carlotta.

'Start at the beginning,' Dr Armstrong said when I made some remark about not knowing where to begin. 'Don't leave anything out. Go on to the end of what you have to say and then stop. Write it as a kind of diary, whenever you feel you have something to say.' He made it sound so easy. Here goes:

19 November 1992

My name is Edward Stonely. I am thirty years old. I teach Art at St Peter's School in the small town of W . . . I'd rather not name it. I am good at what I do. I enjoy the teaching. When I left Art School, I had the mad idea that I might make a living from painting and sculpture, and took the job at the school just to make ends meet till I hit the big time.

That was seven years ago. The big time isn't something I think about any more, not really, though I do sometimes wish I had more time to give to my own work. As it is, I paint and sculpt mostly during the school holidays, and I'm happy to do so, because of Annie and Beth. I would be happy to do almost anything for them, in spite of everything. It's because I don't want to hurt them that I'm writing all this down, trying to get myself sorted out before it's too late.

I have a good job, a lovely wife and child, a house I can afford to pay for, a reasonable future. You would think, wouldn't you, that there was nothing left to wish for, but there is. I wish for peace. I wish I could be rid of Carlotta. There. I've written it down, so now maybe the blackness and solidity of the words will be like a magic spell to wipe my mind clean of those terrible dreams. I feel like Hamlet. I suppose you would say I identify with him, because of what happened to Carlotta, but we studied the play at school and I remember one quotation from it very well because I've thought it myself many times over the past few weeks. 'O God,' Hamlet says, 'I could be bounded in a nutshell and count myself a king of infinite space, were it not that I have bad dreams.' That's exactly what I think.

22 November

I can see it's pointless going any further without writing about Carlotta. I was in love with her for a year when I was seventeen, and she a little younger. She turned up at school one September, out of the blue, and the minute I saw her I knew she was different.

Plenty of other girls in the class were pretty, and I'd had my flirtations with quite a lot of them, kissing in the dark at the movies, or at discos, or parties, sighing a bit but not really suffering when the relationships came to a natural end. Good teenage fun. From the moment Carlotta arrived at school, everything altered for me. The other lads didn't rate her at all, so there wasn't much competition for her favours.

'Funny-looking,' my friend Geoff called her.

'Flat as a pancake,' said Marty, the class sexist pig.

'Silent, too,' said Pete, and their attitudes summed up what everyone thought about Carlotta except me. I could see that her thin body, and her strange, widely-spaced yellowy-green eyes in a somewhat flat face were not conventionally pretty, but they made my mouth dry whenever I looked at her, and my heart pounded when I passed close to her and smelled the wonderful fragrance that seemed to float about her hair.

Her hair . . . even Geoff and Marty and Pete agreed they had never seen anything like it. She wore it long and loose around her shoulders, and it waved and moved as she walked with a life of its own. Everyone called it black, but that was taking the easy way out. I spent hours staring at it, and there were blues and greens and even reds mixed up in the colour somehow, and a gleam on it as it caught the light.

I spent two weeks watching her at the beginning of that term, and then I could bear it no longer. Looking back, I can see that even the way our relationship started was odd. There was no leading up to it, no flirtation, no 'my friend fancies your friend' kind of negotiation that goes on in school romances.

We were in the Art Room, clearing up. There was no one else there. She was washing brushes in the sink. I came up behind her and buried my face in her hair. For a moment it felt as though I were drowning in the fragrance and the softness, and I prayed that I would never ever need to come up for air. She trembled, and then turned to face me.

Have you ever seen dry paper and wood flare up when you drop a lighted match on them? That's what happened to us. To Carlotta and me. Love had set us alight and we caught fire. We crackled and burned and leapt up in blinding flames of scarlet and gold. We were consumed.

For six months, everything seemed to disappear, and there were just the two of us and our passion in the whole world. And then (like a fire) the love on my part began to flicker a little, and dwindle and die. I'm not making excuses for myself. I know what I did was probably harsher than it need have been, but how was I to know that Carlotta would react as she did? I came to the conclusion that our relationship had to end, and I told her so. It happens all the time, doesn't it? Well, doesn't it? Don't boys and girls split up every day of the week with no harm done?

Carlotta seemed very calm when I told her. The yellow eyes

widened. Her face turned quite white. She said a strange thing, one whose meaning I am only now beginning to understand:

'I'm not ready to let you go. Not yet. Not ever.'

Then she turned and left the room and that was the last time anyone saw her alive.

No one had an explanation for how she came to fall off the bridge over the rain-swollen river, with nobody seeing her, nor for why she should have died when we all knew she was a strong swimmer. One theory was that her hair had become entangled on some underwater obstacle as she fell, and that she was unable to free herself. No one else knew that I had ditched her hours before 'the accident'. No one else knew that Carlotta meant to die. Me and Hamlet. Neither of us guessed that love could be so strong, so unforgiving.

29 November

I grieved for Carlotta. Of course I did. I was genuinely sorry that she was dead. Of course I was, but I have to admit that a tiny part of me was furious with her. I can see, I said to myself, what she is saying: 'You killed me, Edward. You did. So suffer.' And I did suffer a bit, but I got over it. I went to Art School. I met Annie. We fell in love and married. Beth was born. I hardly ever thought about Carlotta. Then a few months ago, the dreams started. Dr Armstrong said:

'Tell me what happens in these dreams. Why they are so dreadful.'

'They don't sound dreadful when I tell them,' I said. 'It's Carlotta speaking. Just her head floating in water, with her hair drifting backwards and forwards like seaweed. She says: I'm coming. I haven't forgotten. I'll be there soon. Very soon, and then we'll be together for ever. I shall touch you, she says and then she stretches out two hands in front of her and they're all bones and fragments of skin and I know that if I don't wake up now she will clutch me in her hideous fingers.'

'Hmm,' Dr Armstrong said. 'How very unpleasant. You clearly still feel responsible for Carlotta's unfortunate accident . . . still perceive it as suicide. Have you tried painting a picture of her? Or perhaps making a clay model . . . Maybe that would help . . . giving your nightmares a real, physical presence.'

I promised Dr Armstrong that I would try.

10 February 1993

I haven't written in this notebook for some weeks. I think I may be cured. I have much to be grateful to Dr Armstrong for. The dreams have almost completely left me. Just before Christmas, I started work on a series of paintings I call 'Portraits of Carlotta'. Annie (who knows everything) tried to pretend she didn't mind that I was spending every moment when I was not at school 'locked up in the studio with another woman', as she put it.

All through the Christmas holidays I slaved over my canvases. There are enough here for an exhibition, but I am reluctant to let anyone see them. Carlotta's yellow eyes follow me wherever I go. There's one portrait in particular I'm pleased with, where she seems almost to be walking out of the frame and into the room. She has her hands held out in front of her like a sleepwalker. Sometimes, I find myself wanting to touch her, and I put my hands out so that they almost reach the hands in the painting.

20 February

How am I going to write this? I pray that my darling Annie will never read it. But I have to say it. If I put it down on paper, it may not be so dangerous. If I don't say it, I feel the strength of my own feelings may cause me to explode. A strange thing happened last night. I had nearly finished another portrait of Carlotta: a close-up of her face taking up almost the entire canvas. I was painting the half-open mouth when suddenly I found that the brush had fallen out of my hand and my own lips were touching Carlotta's. I was kissing her image. This was bad enough. Worse, oh, so much worse, was what I was experiencing as my mouth came into contact with cold paint. I was as stirred by this lifeless kiss as I had been the first time I had kissed the real Carlotta, all those years ago in the Art Room.

I tore myself away from the painting, and went to lie on the studio sofa, feeling sick, and hot, as though some dreadful fever had seized hold of me. I feel somewhat calmer now, writing this away from the studio, where she can't see me, but now that I am calmer, I can face the truth. She has enchanted me all over again. I want her. I wish we could be together once again. I think she has driven me mad.

Yesterday I found myself looking at her outstretched hands in the portrait I like best of all. The perspective has worked, I thought. She really does look as though she is about to walk out of the frame. I put my hands out and touched her painted ones. This must be what an electric shock feels like, I thought. I must stop. I must stop this madness now before it's too late. Oh, Carlotta, I am longing, hurting, burning for you! I cannot bear it.

10 April

The exhibition is over. It was an enormous success. Every single one of the portraits has been sold. The local paper called the show 'a moving tribute by one of her old schoolfriends to the tragic victim of a drowning accident'. I am a celebrity at school. Annie, I can see, is mightily relieved to have the canvases out of the house. I am bereft. I cannot live without her. I shall make her again. Differently this time. I shall take clay and fashion her once more. I shall paint the clay in all the colours of her skin and hair . . . bring her to some kind of life. I shall tell no one at all about this. It will be our secret. Carlotta's and mine.

1 May

It is done. Now we can be together. Carlotta, fetch me. Take me with you.

* * *

3 May 1993 Extract from the statement of Mrs Anne Stonely, submitted in evidence at the Coroner's Inquest

Beth was fast asleep in her room. We always had our supper, Edward and I, after she had gone to bed. I called him to come and eat, but he didn't come, so I went to find him. There was nothing strange about this. He often became so absorbed in his work that he lost all track of time and place.

I went into the studio. I hadn't been in there for ages. I didn't know he had begun to work in clay. As soon as I saw him, slumped like that against the model of a woman, I knew he was dead. His legs were trailing on the floor. Her arms were around him. It was almost as if she were holding him up. His head was thrown back. The woman's head had lolled forward. It was horrible. They were tangled up together. The worst thing of all was the hair. Her hair seemed to be filling his mouth. It almost looked as though he were eating it. I became hysterical. I knew who she was supposed to be. I recognized the long black hair. It was Carlotta. I ran out of the studio screaming and my neighbour heard me, and phoned the police and our family doctor, Dr Cooper.

Later, Dr Cooper told me what he had found. I can't understand it. Edward, it seems, had choked to death, even though they had no idea at all of what might have choked him. There was no sign at all of anyone else having been in the studio, although the floor was awash with water. No one knew where this had come from. Neither Dr Cooper nor the police could find any sign of a woman fashioned out of clay. They took me to the studio to show me, and I couldn't see her either. Not any more.

Dr Cooper says I must have become hysterical when I discovered Edward lying there dead, and imagined the whole thing. I must have. I hope that I did. Otherwise, what happened, and where Carlotta is now, are too horrible even to think about.

Little Black Pies

JOHN GORDON

Ghosts,' she said. 'There ain't no such thing.'

Emma Stittle watched her plump arm spread and become fatter as she pressed it on the kitchen table.

'There ain't no such thing,' she repeated, and with her thumb stripped peas from a pod, cupped them in the palm of her hand and raised them to her mouth. 'Ghosts is a load of old squit.'

Her sister Sarah, thinner and older, rattled the poker between the bars of the kitchen stove, and red coals and ashes fell into the grate. 'It get so hot in this kitchen on a summer's day,' she said, 'that I wonder I bother to cook anything at all.'

'Me,' said Emma, 'I reckon it's stupid to think that people who eat can die and then turn into somethin' that don't need food.' She chewed as she slit another pod with her thumbnail. 'I like peas, but there ain't no substance in 'em.'

Sarah had no time to talk of ghosts. 'I'm that hot I don't know what to do with myself,' she said as she hobbled towards the cottage door, narrow-shouldered and stooping. She lifted the latch and pulled the door open. 'Come you on in then,' she said. 'All on you, you little black devils.' A great ramp of sunlight streamed through the door to brighten the tiled floor. 'Flies,' she said, 'swarms on 'em.'

There was a dance of black specks above Emma's head but she paid no heed. 'When our old mother used to talk of ghosts,' she said, 'she used to make me frit. She used to say they come back because their time weren't properly run. That there was unfinished business or somethin' like that. Load of old squit.'

'Who does all the work round here? Who?' Sarah, like her sister, wore a flowered, wrap-around apron, and both had their hair drawn back into buns, but Sarah's hair was grey while her sister's was still glossy black. 'I scrimp and save and slave and scrub,' she said, coming back to the table, 'and what thanks did I ever get for it? She died, didn't she? And never no word of thanks. Not one.' She banged an earthenware bowl on to the bare wood of the table in front of Emma and began scooping flour into it from a big jar.

Emma sat where she was, her round cheeks dimpling as she munched. 'You look a bit wore up today, Sarah,' she said. 'What's been gettin' you down, gal?'

Sarah cut a wedge of lard and began rubbing it into the flour with her skinny brown fingers. 'I slaved all o' them years after Father died and Mother was left alone. I skivvied from morning till night and what help did I get?' She dug savagely into the white mess.

'I were much younger than you, Sarah, don't forget. Only a little kid when Mother were taken poorly. And I used to sit beside her up in her room and keep her company hour after hour.'

Sarah did not look her way. Her blue eyes, fading with age, gazed at where her fingers dug. 'Spoilt brat.' The wrinkles of her thin face arranged themselves into a simper and her voice took on an acid whine. 'Please, Sarah, Mummy's sent me down for a cup of tea. I'll take it up, Sarah. Mummy don't want nobody else to disturb her.'

'I weren't ever as bad as that. You're putting it on.' Emma was laughing. 'But she did like me to sit with her. She used to sit in that old high-backed chair by the window, with her shawl around her, and her pillows, and I used to pull aside the lace curtain so as we could see down into the lane, and she used to tell me things about everybody that came by.' Emma chewed and laughed again. 'She told me things no kid ought never to have been told. About women in the village. And men.'

Sarah paid her no attention. She poured a little water from a jug into the basin and mixed it in with a knife, jabbing. 'Best frock. Always best frock because, "Mummy likes to see me pretty." And who done the scrubbing with a sack tied round her waist? Who took out the ashes and blackleaded the stove? Twice a week I done that, and all the cracks in the skin o' me fingers shown up like black spider webs.'

But Emma was in a reverie. She gazed through the open door across the lane to where the yew trees shaped themselves against the blue sky. 'She used to love a funeral. Especially if they dug the grave near enough so she could see the coffin go down. She used to love telling me how people died. Little Claudie Copp called for his Mum all through one night, she say, and he never once see her alongside his bed. And then in the morning, Mother used to say, he went up to heaven with the angels.'

'Like black spider webs.' Sarah slashed the dough across and lifted half of it on to a piece of oilcloth where she had sprinkled flour. 'I could've had nice hands. Nice soft white hands like I seen some girls have, but they was always in water. Always scrubbing.'

'It was lucky for her we lived just across from the churchyard,' said Emma. 'Gave her an interest. People always visitin' graveyards. And stayin' there in the end.' She laughed again and looked towards her sister, but Sarah had a new grievance.

'I could've had him if I wasn't so thin and dry and wore out with work. I could've.' She rubbed flour on to the rolling pin and began to roll out the dough. 'I had his child, didn't I?' The pastry was a thin white island on the oilcloth. 'I had his little baby.' A tear came from the inner corner of one faded eye, but was so thin it did no more than moisten the side of her nose.

'You never did!' Emma's eyes gleamed with surprise and curiosity. 'Whose baby? You never said nothing. I never seen no baby.'

Sarah had turned her back to get a pie dish from the window ledge. 'I never told nobody. Nobody ever knew.'

'But who was he, Sarah? I never knew you had a feller.' Her sister remained silent, and Emma became sly. 'I don't mean you never had admirers. I used to think my Tom looked at you a bit. Used to, till I made him stop.'

Sarah lifted the pastry and began to line the pie dish. 'Tom were a lovely man.' The glisten of a tear came and went. 'He were lovely.'

'So I were right.' Emma turned her plump face away and patted the glossy bun at the back of her head. 'I guessed as much. Tom never said nothing but I guessed as much.'

'If I'd told him about the child he would have married me. I

know that. But I could never hold a man because of that—not
when his eyes was on my own sister.'

Emma had jerked towards her, her mouth open, but Sarah's
voice did not change its pitch.

'So I lost him, didn't I?' She had lifted the pie dish and was
trimming the pastry at the edge. 'And there was nothing but the
baby.'

'Tom's baby.' Emma's whisper was as soft as the ash that fell
in the grate.

'It died,' said Sarah. 'I made it die. And it lies yonder still,
under that tree. No father, no mother, nothing.'

Emma's round face was pinched suddenly and her voice was
harsh and vindictive. 'It's as well for you there ain't no such
thing as ghosts, Sarah Stittle, or else you would be haunted!'

But Sarah spoke as though Emma was not there. 'First she
took Mother from me, then she took Tom, and I never said a
word. Never said, and I ought to have done. I ought to have
said.'

Emma opened her mouth to speak, but a sudden flutter of
wings in the doorway made both sisters start. A jackdaw,
twisting his grey nape in the sunshine, stood on the step.

Emma gasped. 'My God, that were like the angel o' death. That whole doorway seemed full of black wings. I can't stand birds. I hate them stiff feathers. Go away! Get out, you devil of hell!'

But Sarah was wiping her hands on her apron as she went towards the doorway. Her face softened with pleasure. 'Come on then, my beauty. Come you on in and see your Auntie Sarie. There now, there now.' She stooped and held out a finger. The bird hopped on it. 'Lovely little cold black claws you've got, my lovely boy. Hold tight to your auntie.'

'Take it away! Take it away!' Emma shrank back in her chair. 'Please, Sarah!'

But Sarah spoke only to the bird. 'You came when I needed you, my lovely. You came hopping over the road just when I were down in the dumps, my lovely boy.' She raised the bird and her dry lips touched his black bill.

'Sarah! I can't bear it!'

'Just when I needed you, you came hopping along with your black eye. And didn't you know it all, didn't you just know it?' For the first time, the bird against her cheek, its black feathers touching her grey hair, she looked directly towards her sister.

'Emma took my man, didn't she? Emma took my man. But you showed me, didn't you, boy? You showed me how he'd never have her. Never no more. Skippety along the lane, skippety down the hollow. You showed me the pretty flower and the little black berry. The little black berry I put in the pie. One, two, three . . . many, many more. In a little pie for Emma. Emma's little pie.'

'You're talking daft, Sarah. What you on about? Throw that bird out. Get rid of it.'

'And Emma never knew.' Sarah sat and stroked the bird, looking no more towards her sister. 'Emma never knew about them little black berries what she ate. Ate many and many a time.'

'My stomach,' said Emma. 'I had a bad stomach and you gave me little pies to ease it. They was nice.' She tried to smile, but although the dimples came they were pale.

'I gave her pies, my beauty. Little black pies. And now she ain't got no stomach-ache no more. Nothing no more.'

Emma made herself laugh. 'Sarah!' she called. 'Sarah, look at me!' But Sarah did not stir. 'Sarah, you make me feel bad. What did you do to me?'

Sarah held the bird so her nose was touching its deep grey cap. 'I wish we could tell her what we done, Jack my beauty. I wish we could tell her, but it's too late now.'

'What do you mean too late? What you done to me?'

But Sarah ignored her sister's cry, kissed the bird and put it down on the table. 'You like to peck peas, Jack. There you are, my little boy, my lovey. Go you peck them peas.'

The fat woman pressed her arm on the table and clenched her fist on the peas.

'You ain't going to scare me,' she said. 'You ain't going to scare me with your talk. You talk as if I was dead. But if I was, how come I'm here?' She laughed, defying her sister. 'There ain't no such thing as ghosts.'

The bird's black claws skittered on the table top as it went towards the clenched fist. Emma clutched tight, refusing to move. The bird stabbed down. She clutched tighter and shrieked. But the room was silent. And the black beak pecked through a hand that nobody but Emma herself could see.

The Guitarist

GRACE HALLWORTH

Joe was always in demand for the Singings, or community evenings held in villages which were too far away from the city to enjoy its attractions. He was an excellent guitarist and when he wasn't performing on his own, he accompanied the singers and dancers who also attended the Singing.

After a Singing someone was sure to offer Joe a lift back to his village but on one occasion he found himself stranded miles away from his home with no choice but to set out on foot. It was a dark night and there wasn't a soul to be seen on the road, not even a cat or a dog, so Joe began to strum his guitar to hearten himself for the lonely journey ahead.

Joe had heard many stories about strange things seen at night on that road but he told himself that most of the people who related these stories had been drinking heavily. All the same, as he came to a crossroad known to be the haunt of Lajables and other restless spirits, he strummed his guitar loudly to drown the rising clamour of fearful thoughts in his head. In the quiet of early morning the tune was sharp and strong, and Joe began to move to the rhythm; but all the while his eyes were fixed on a point ahead of him where four roads met. The nearer he got, the more convinced he was that someone was standing in the middle of the road. He hoped with all his heart that he was wrong and that the shape was only a shadow cast by an overhanging tree.

The man stood so still he might have been a statue, and it was only when Joe was within arm's length of the figure that he saw any sign of life. The man was quite tall, and so thin that his clothes hung on him as though they were thrown over a wire frame. There was a musty smell about them. It was too dark to see who the man was or what he looked like, and when he spoke his voice had a rasp to it which set Joe's teeth on edge.

'You play a real fine guitar for a youngster,' said the man, falling into step beside Joe.

Just a little while before, Joe would have given anything to meet another human being but somehow he was not keen to have this man as a companion. Nevertheless his motto was 'Better to be safe than sorry' so he was as polite as his unease would allow.

'It's nothing special, but I like to keep my hand in. What about you, man? Can you play guitar too?' asked Joe.

'Let me try your guitar and we'll see if I can match you,' replied the man. Joe handed over his guitar and the man began to play so gently and softly that Joe had to listen closely to hear the tune. He had never heard such a mournful air. But soon the music changed, the tune became wild and the rhythm fast and there was a harshness about it which drew a response from every nerve in Joe's body. Suddenly there was a new tone and mood and the music became light and enchanting. Joe felt as if he were borne in the air like a blown-up balloon. He was floating on a current of music and would follow it to the ends of the earth and beyond.

And then the music stopped. Joe came down to earth with a shock as he realized that he was standing in front of his house. The night clouds were slowly dispersing. The man handed the guitar back to Joe who was still dazed.

'Man, that was guitar music like I never heard in this world before,' said Joe.

'True?' said the man. 'You should have heard me when I was alive!'

The Chocolate Ghost

JULIA HAWKES-MOORE

The friends Kate and Sarah made chocolates. In fact, they looked rather like the chocolates which they made. Both girls were very large, tall and wide, plump and bosomy, with skins as glossy as cocoa butter. Their long shining hair was swept up with ribbons tied on top. They wore frilly aprons with lacy petticoats peeping out like decorative doilies under the hems of their flounced skirts.

Kate's hair was plain-chocolate dark and Sarah was milky-blonde. Kate was as tough as Brazil nuts beneath her crispy outer shell, whilst Sarah was sugar sweet right through to her soft centre. Their smiles were as toothy and welcoming as those on their Easter white chocolate bunnies, and they were altogether as delightful as the luscious plains, milks, and fudges which they made each week.

The chocolate girls decided to buy a shop. They only had a little money, so they had to hunt around for a long time before they found somewhere cheap and pretty enough to suit their purpose.

In a little cobbled street just off the market square they found a tall, narrow house squeezed in between two strict Georgian red-brick ones. It was all tottery with carved oak beams, and diamondy window-panes. It was cheap because it had been empty for many years and the pre-war paint had all peeled away. It had once been a sweetshop, but no one had wanted to move into it, because it was rumoured to be haunted.

Certainly the place had an atmosphere, which very slightly raised the hair on the back of their necks as the girls walked around with the estate agent. But nothing could bother the chocolate girls. Merry, practical spirits, they shook hands on the agreement then and there, tripping gaily away to sign the deeds.

Surprisingly soon they were back, proud as princesses of their new realm. They whooped for joy as the door swung open to their key, and danced in the dust. They attacked the huge task with gusto, and half-filled their hired skip with debris before darkness fell on their first day of ownership.

Tired and filthy, they slammed the front door with a happy thud. Gazing up at the scruffy but pretty building which was now theirs, their eyes caught a slight movement in the topmost window. It looked for a moment as though a body swung by its neck behind the cobwebbed panes.

'Trick of the light,' Kate decided, and they sauntered away.

Next morning, they were back early to start again. To their surprise, the neatly stacked brooms, mops, and pails lay sprawled across the floor.

'We must stop slamming that door!' laughed Kate, and they picked up all the things and went back to work.

At last the shop was hygienic and sparkling, and they decorated it in rich milky creams and toothsome dark chocolate browns. It was all frilled with sweeping curtains, lacy nets, and golden ribbons, the entire shop a beautiful box in which to display their finest chocolates. There was a grand opening, involving the Mayor and Mayoress, champagne and rum truffles. The gleaming old inlaid brass cash register was soon tinkling merrily with the chocolate girls' incomes.

Yet as Sarah worked, she became increasingly aware of the feeling of being watched; but every time she turned around sharply, she saw nothing but shadows.

Late that evening, Kate studied Sarah's frowning face under her fluffy blonde fringe. 'It's not like you to get so worried about something, Sarah. What is it?'

'Kate, you don't suppose that there's a ghost in the shop, do you? It's just that I'm starting to get the feeling that we're not very welcome. There is something nasty here, I don't know what. But I do know that it doesn't like us one bit. I feel . . . scared of it.' Sarah blushed foolishly under her friend's stare. 'Just forget it, I'm talking nonsense.'

Kate continued to look at her silently. Although she was unwilling to admit it, she too had become gradually aware of small sounds, of flickers of movement in the corner of her eye, and the hair stirred on the back of her neck. She reached up to rub the suddenly aching muscles of her shoulders, and yawned.

'Ghosts, indeed! I don't know about any daft old ghosts. All I know about is making chocolates, and we've got a week's supply of cream truffles to make tomorrow morning. We're just overtired after all our hard work. Tell you what, let's go down to the Three Crowns tomorrow night; there's a good band on. We'll get all tarted up, and check out the talent, hey? It's about time we treated ourselves to a good flirt with some dishy fellas!'

Sarah giggled at her suggestion; 'Will do!' she cried. 'I'll wear my new red Lycra dress. Now go and get some sleep, Kate; you look ready to drop.'

Kate yawned again, and turned away. She began to climb the stairs slowly and heavily, watching her feet on the narrow steps. Sarah glanced up towards Kate's bedroom door, then froze in horror. She let out a shriek:

'Kate! Behind you, Kate, look!'

Kate glanced up the last few stairs and cried out in disbelief. Staring down at her, just inches in front of her own face, was the wild-eyed and vicious face of an old man. His hair stood out in an unkempt and greying shock, his stained teeth were bared in a wide snarl. Although his form was smoky and insubstantial, Kate, rigid with fear, felt the clutch of his pincer-like fingers on her shoulders, and tried to scream out for help as he spun her round, then gave her a mighty shove. The scream only began to spill out of her open mouth as she fell headlong down the steep staircase. She stretched out her arms, trying desperately to slow her heavy fall.

Sarah rushed forward to catch her friend as she reached the bottom of the stairs, and the two girls collapsed in a heap. As her knees buckled beneath Kate's weight, Sarah glanced upwards, only to see a smoky haze fading away into nothing, but leaving the distinct impression of a leering face, laughing down at the disaster below.

Kate moaned, stirred and burst into tears, Sarah joining her a split-second later, as she helped Kate to uncoil and straighten up. Then Sarah flung her arms around Kate's neck, and they both cried and cried. The horror of the hideous face which they had both seen, the shock of Kate's fall, and the tensions and worries

of the last few weeks all flooded out, until their hair and faces were wet and their eyelids swollen.

At last, all cried-out, the girls separated, to catch their breath and look at each other, wide-eyed with disbelief at the situation. They each glanced back upstairs, but Kate shook her head fiercely.

'No way am I staying here tonight. You can, if you like, but I'm not going up those stairs again in a hurry. I'm going back to my parents' place tonight. Get your stuff, if you want to come with me.' Kate's words were clear, but her voice shook with emotion.

Sarah nodded, and hurried to collect her coat and bag, then they carefully descended the next flight of stairs into the shop. They glanced around, sadly, at the display of hard work and hope which they had put into the shop, both sighing. Then they unlocked the shop door, closing it gently behind them.

'Burglars?' Sarah enquired tearfully, the next morning, as the girls gazed around them at the debris of a once-lovely sweet shop.

The glass shelves of beribboned boxes of hand-made chocolates had been swept clear, and all those pretty things lay squashed and mangled on the floor.

Kate pointed at the walls, jabbing her finger at the ugly scrawls of loopy writing, smeared across the creamy paintwork. 'What about all that?' she demanded. 'Obscenities and filthy suggestions. Not what we want our customers to read. No, Sarah, that's not burglars, not with twenty quid still in the till, and none of the selection boxes missing. That's the flipping supernatural. You were right. We are haunted.'

'Can we get it exorcized?' Sarah sobbed.

Kate drew a deep breath. Thrusting a tissue into Sarah's hand, she declared, 'Right! Let's go and see that snooty estate agent who sold us the shop. We'll sort this out!'

The estate agent, confronted by Kate, sighed, and told them the story of the haunted sweetshop.

Long ago, it had been bought by a young man called Jim Evans. He had planned to share it as a new home and business with his fiancée. She was a pretty little thing, all pouts and ringletty curls, and with a very sweet tooth. Jim courted her with barley sugar, pressed her with candyfloss, and finally won her hand with a bagful of marrons glacés. Tragically, the war broke out during their gentle and sticky courtship, and sugar rationing ruined his hopes. He bribed her with sugared almonds for kisses, and then he couldn't supply enough sugared almonds.

She resisted his increasingly shabby advances throughout the war, until all that he could offer was the promise of marzipan on their dried-egg wedding cake. She jilted him when he confessed that he could not afford the icing-sugar, and then she ran off with an American GI, who enticed her with unlimited Hershey bars.

Her abandoned lover hardened his heart against all women, with the ferocity of boiling toffee. He short-measured any female over the age of nine (or younger if they wore ringlets), and was so surly to the mothers and grandmothers of his little customers that his trade fell slowly but surely away.

Eventually, Jim Evans found himself without any income or friends, unable to pay his rates demands, and heartbroken, all because of that long-lost young lady with a sweet tooth.

One dark night, Jim took a rope and hanged himself in his bedroom, but his body spun there slowly, undiscovered for days. No one attended the funeral.

'What a sad, sad story,' whispered Sarah, as they trudged home. 'No wonder Jim Evans hated us young women moving into his house and selling sweets in his shop. It makes me want to help him in some way, not just banish the poor man . . .'

Kate regarded her thoughtfully. 'Jim Evans didn't just sell sweets, either—he used to make his own sweets like we make our own chocolates, and jolly good they were, too, that estate agent said. You can't get those flavours today . . .'

'I beg your pardon,' said Sarah indignantly, wiping away her tears. 'It's the big businesses that have changed all the techniques, but you and I can recapture the old taste of sweets, you know, because we work on such a small scale. It's just the recipes that are hard to find.'

Kate turned to look at her in surprise. Sarah pointed to the scrawls on the wall. 'And if Jim Evans can write rude words— even if he can't spell them—then he could write his recipes down as well. We could do a deal with him.'

'Are you suggesting that if we don't exorcize him, he might give us his recipes?'

Sarah's eyes were gleaming, as she seized Kate's hand; 'Kate, don't you see? He doesn't mind us selling sweets here, but he gets upset whenever we talk about meeting men. He's jealous and frightened because we might run away with other men, like his fiancée did. But if we work very hard, as we ought to, setting up a new business, then he's on our side. He could be a sort of sleeping partner, helping to build up the business of his dreams. But we must be faithful to him, if he's going to give us his precious recipes. I will, anyway! I want to make a go of this business. What about you, Kate?'

Kate stared. 'You're right, Sarah. I'll do it.' She shouted into the air: 'Jim Evans, we're the girls for you! Here's the deal. You give us your recipes, we'll sell them. And we'll be good girls. We'll look after you, if you look after us.' She started to laugh, and, grinning, sprang over to the gilt-edged blackboard where the flavour and price of the weekly special truffles were chalked up, and wiped it clear with her sleeve. She rested a small stick of chalk against it, and stepped back.

Then she grabbed Sarah by the wrist, pushed her into the kitchen at the back of the shop, and firmly closed the door behind them both.

'Now,' she declared, 'we give him his chance. Sarah, the broom. Let's start tidying up in here—we've got a business to run.' She grabbed the mop, and started feverishly washing down the walls.

The kitchen was restored to sanity and hygiene long before they dared to venture back into the shop. Everything was as disordered and tangled as before. With great trepidation, they tiptoed over the carpet of squashed sweets towards the blackboard, and peered at it in disbelief.

Covering its surface with close-written, misspelled writing was a recipe.

'Crystallized flowerpetals,' Sarah read aloud. 'Jim is sorry; he does want to help. And he's giving us bouquets of flowers to apologize, in the only way he knows how. Oh, you sweetie-pie!' she trilled to the empty air. 'Thank you. We'll make them at once. They'll sell really well, won't they, Kate?' She pirouetted around to look into the face of her friend, her eyes alight with hope.

Kate regarded her silently, and bit her lip as she considered. 'Well,' she mused aloud, 'we could try making them this evening, instead of going to that dance, to see if the recipe really works.'

Sarah clapped her hands in glee. 'Yes. Oh, you lovely man, Jim Evans!' she shouted triumphantly into the air. 'This could be the start of something really special, don't you think, Kate?'

Kate smiled and hugged her friend. Both girls then turned and began restoring the shop to order.

Soon, the sugar-kettles in the kitchen simmered and steamed, and the air was treacly and fragrant with flowers and caramel. Kate plucked petals into china basins. Sarah dipped and dried and carefully arranged.

The completed articles were enchanting. The boxes were laid out in formal swirls and lozenges like the knot-gardens of medieval manor-houses. Strips of succulent green angelica were interwoven with lengths of barley-sugar around panels of delicate rose-petals, vivid marigolds, and fragile preserved violets. Stepping-stones of crystallized ginger lay across pools of blue candied lavender buds. Little mounds of sugared coffee beans appeared like tiny rockeries, garlanded with pearls of coriander seed and fronds of maidenhair fern, frozen beneath a crackling glaze of syrup.

Made with love and lavished with attention, these confections were bought as soon as they appeared for sale. The shop became so busy that they had to expand the shop-space into the kitchens, and hire workshops in a nearby village.

Jim Evans seemed to have repented fully of his cruelty to the chocolate girls when they first opened their shop in his home. For Jim Evans, failure, outcast, and reject of society as he had been, to have the world rushing to his door to buy his confectionery pleased him deeply. The atmosphere of gloom within the house had evaporated in the bubbly, frilly and merry charm of success which seemed to bless every new venture of the chocolate girls' business empire. Kate and Sarah greeted Jim with thanks each morning, as a new recipe appeared. They wished him farewell each night, leaving out samples of the new sweets, which had always vanished by dawn.

The public bought the sweets, and were consumed by memories. For elderly people, a single cachou or parma violet brought images of lost and beloved childhood faces. Younger people explored the mysteries of jujubes, lozenges, liquorice laces, and aniseed balls. Tigernuts, Spanish tobacco, bulls' eyes, and sherbet suckers were bought by the pound, whilst jamboree bags with all their hidden surprises proved very popular. Love-letters and satin cushions were exchanged by lovers as frequently as were the original truffles and hand-made chocolates which Kate and Sarah still produced.

At last the morning dawned when Jim had no more recipes to leave for his chocolate girls. His repertoire was exhausted, even to its last sugar mouse. Ready for work, the girls descended to find a tender farewell blessing chalked onto the blackboard, signed with two large wavering kisses. As they stood still in surprise before the board, they each felt the touch of a hand laid gently on their shoulders. Turning slowly, Kate and Sarah looked into the sadly smiling face of an old grey man standing silently behind them. He leaned forward, wisping a kiss over the cheek of each girl, softly as a cobweb. Then the ghost of Jim Evans faded away for ever.

Sarah wept copiously, and even Kate had to brush away several tears. They hugged each other to comfort their sudden sense of loss.

The shop seemed bigger and quieter without Jim, and the future seemed blander. But not for long.

Sarah wiped her eyes and gazed up at Kate. A slow, curling smile broke across her tear-streaked face, and her wet eyes gleamed in excitement. 'Well, Kate, now that Jim has left the shop, you know what we can do at last, don't you?'

Kate looked at her, and a beaming smile gradually lit up her features. 'Yes, of course! Now we can really start to live—it's been all work and no play for quite long enough, eh, Sarah? We've been very good for a long time, haven't we? Now let's start celebrating our success, by going to a dance!'

As for the generous spirit who had inspired and dictated their success, Kate and Sarah immortalized him in the name of their traditional confectionery company: *The Chocolate Ghost*.

Children
— *on the* —
Bridge

KENNETH IRELAND

M r Fulwood was delighted with the village in which he had bought a cottage for his retirement. He had always fancied living in such a pretty place, had dreamed of a village just like this during his whole working life, out of the hubbub and hurry of the town. Now that he had retired, his dream, he was sure, had come true.

Mr Fulwood was a bachelor, well used to managing by himself. With his pension from work, plus his State pension, he would be able to manage very nicely, and the cottage had come up for sale at just the right time. Not that the village was isolated—he would not have liked that—because it had a regular bus service into the nearest market town, and that was part of the attraction. He was now able to live in the countryside, yet within a bus ride of the conveniences of a town. Just the right sort of combination.

And the village really was pretty. It had a shallow river running through it with meadows on either side, at one point passing underneath a little bridge which carried the main road through the village; it had an ancient church with old houses clustered round it as if for protection; there was a scattering of farms; a real old-fashioned village inn; a general village shop and friendly people. What more could he want?

Even the children were friendly, far different from the preoccupied, surly ones he had encountered living in towns. He noticed some on his first morning out, down to the village shop. They were sitting on the low wall of the bridge as he passed, five of them he counted, all looking clean, fresh and bright.

'Good morning,' said the first of them.

'Good morning,' he responded cheerily, then they all called the same back to him again . . . not shouted after him, but just called a greeting.

On his way back from the shop they were still sitting there in a row, all about the same age it seemed to him, except that the tallest one who sat at the end, who had spoken to him first, was perhaps a year older or thereabouts.

'Got everything you wanted?' the nearest to him asked, noticing his carrier bag.

'Yes, thanks,' said Mr Fulwood, and smiled at all of them.

'Be seeing you,' he heard as he passed.

Every day he made a point of going to the shop for some small thing, at first because the short walk was good for him— the fact that he had retired was no excuse for just pottering around his cottage in his bedroom slippers, he had decided. But then he had another motive, which was to have the pleasure of seeing these delightful children.

They would either be sitting on the wall and swinging their legs, or they would be down by the river. Or if the day was warm, they would be playing in the river with their shoes and socks off. They were all boys, he noticed; not a girl among them. But that was understandable, he supposed, because at their age boys and girls tended to play separately. He knew that was how it was when he had been their age. Every time they saw him, they would either call out to him, or if they were down by the river they would turn and wave, and he would wave back. He began to enjoy their company.

'Come down and join us,' the tallest one called to Mr Fulwood one afternoon. They were down at the river-side.

'How do I get down?' he asked.

'Just climb over the fence at the side of the bridge, and then come down. It's easy.'

He had not climbed a fence for a good thirty years, but he found that this low wooden one presented no problem. He smiled to himself, at the thought of what his friends in the city would say if they could see him now, not only climbing over a fence obviously placed there to keep people out, but in his oldest clothes as well. In the city, just a few weeks before, he would never have dreamed of appearing in public in anything but a suit. This country life was just the thing, he was thinking, as he almost

ran down the grass of the meadow to where these boys were waiting for him.

'This is Charlie,' said the tallest boy, introducing that one to him, 'and this is Trevor, and this is Mark, and this is Darren, and I'm Errol.'

'How do you do?' returned Mr Fulwood politely. 'I'm Mr Fulwood.'

'We know,' said Charlie. 'We know the names of everybody in the village.'

'Of course you would,' said Mr Fulwood. In such a small village, everyone would know everybody else. He was expecting to discover the names of most of the other people living in the village before long. He was learning every time he called into the village shop, because the shopkeeper always addressed the customers by name.

All the boys had their shoes and socks off.

'Do you want to come in with us?' asked Trevor.

'In where with you?' asked Mr Fulwood jovially.

'The water, of course,' said Errol. 'You can if you like. It's not very deep. There are only tiddlers in there, but you can catch one in your hands if you're careful. You only have to paddle.'

'Paddle? At my age?' Mr Fulwood was amused at the thought.

'Why not?' demanded Darren. He was the smallest, with a cheeky face. He was the one who had once asked if Mr Fulwood had everything that he wanted.

'All right,' said Mr Fulwood after a moment's hesitation. 'I will! Why not?'

He took his shoes and socks off, rolled his trousers up to the knee, and followed them into the water, which felt delightfully cool to his feet and ankles.

'We sometimes go right in,' confided Darren, 'but not when anyone's around. We've got no trunks, you see.'

'We've got no trunks when we decide to go right in,' corrected Errol. Mr Fulwood could see no reason for the correction.

'That's right. Not when we decide to go right in. You wouldn't want to join us then.'

'No, I think I'd better not then,' said Mr Fulwood hastily.

'Show you how to catch a tiddler,' said the one who had been called Charlie.

Mr Fulwood had a thoroughly enjoyable time trying to catch tiddlers. He didn't quite manage it, and once he almost fell full-length in the attempt, to be rescued immediately by two or three of the boys who seemed to be keeping a careful watch on him.

To be honest with himself, he did not see why these children should want to be friendly with an elderly man like himself. He had to admit that he would not have selected himself for company at their age. But perhaps it was part of their nature to be friendly, he decided. They recognized him as one who needed a few friends, no doubt, he having just moved in amongst strangers, and so were doing their bit to make him feel at home in the best way they knew. And very pleasant it was too, he also had to admit.

Then he looked at his watch.

'Good heavens,' he said, 'it's well past tea-time and I ought to be going.'

'Aw, you don't have to,' said Trevor.

'Oh, yes I do,' said Mr Fulwood. 'Um . . .' he looked around. 'How do I dry my feet?'

He had forgotten all about having to do that.

'We use our hankies,' said Errol.

'Good idea,' said Mr Fulwood, and sat on the grass, took out his handkerchief and dried himself properly before putting on his shoes and socks again.

'Come and join us—any time,' one of them called as he walked back up to the fence, climbed over and set off along the road for home.

'I will,' replied Mr Fulwood.

Every day when it was fine the children were there by the bridge over the river. It took a long time for him to realize something.

'Don't you go to school?' he asked suddenly one morning, when he was just sitting on the grass watching them play.

'Not now,' said Mark. He was the one who spoke least as a rule.

'Holidays, is it?'

'Sort of,' said Mark.

They were playing tig at the time, so at that point he had to run off again.

'Do you want to play tig, Mr Fulwood?' asked Charlie.

'At my age?' he laughed.

'Can if you want to. We enjoy having you with us.'

So he did play tig, and the next time he went down there he took along a tennis ball which he had bought in the village specially for the occasion and taught them how to play hot rice. It seemed they didn't know how to play that, but he remembered how from years ago, when he had been at school. It was surprising how quickly it all came back to him.

The road through the village was never very busy, but every now and then one or other of the villagers would walk past and look at Mr Fulwood curiously as he sat on the wall or played with the boys on the meadow, or waded about in the shallow river. They would never say anything to him when he was there. Probably they thought he was a rather strange old man to be playing with little kids at his age, but he didn't mind. He was happy.

One did say something to him once about it, actually. It was one of the really old men who lived in the village, hobbling along with the aid of a stick past his cottage as Mr Fulwood was just emerging from his gate.

'Morning, Mr Fulwood,' said the old man, then stopped.

'Good morning, Mr Cooper,' greeted the other breezily.

Mr Cooper, however, stood still, leaning on his stick, instead of walking on. 'I ought to tell you,' he said slowly, 'because many round here won't. You ought to keep away from that bridge.'

'Oh, that's all right, Mr Cooper, it's perfectly innocent,' replied Mr Fulwood.

'That may be,' said the old man solemnly, 'but you ought to take warning before it's too late.'

Then he hobbled on towards his own house further along the road, and would say no more. Indignantly Mr Fulwood set off towards the village, waving to the boys who were clearly expecting him to pass, and telling them that he would be back presently, after he had been to the shop.

Mrs Guffy, the shopkeeper, was for once alone in the shop when he entered. He was still rather disturbed at old man Cooper's comments about those innocent children. It had almost sounded as if he had no liking for them at all, and he couldn't understand such an attitude. He looked around the shop after he had made his small purchases, and his eyes lighted on some bags of boiled sweets.

'And I'll have a bag of those,' he added, 'and the boys can share them among themselves.'

'What boys?' asked Mrs Guffy sharply, looking up.

'Errol and Charlie and the other boys I see every time I come across the bridge,' he explained. 'Is there something wrong?' For Mrs Guffy's face was definitely registering almost horror of some kind.

'You've been with them?' she asked.

'Well—yes.'

'You've spoken to them?'

'I chat with them quite a lot,' he said. 'They seem to like me to join them.' He didn't like to say that he actually played games with them, but Mrs Guffy was already ahead of him.

'So that's why you bought that ball the other day,' she said.

'Why, yes.'

Mrs Guffy leaned forward over the counter earnestly. 'Mr Fulwood, don't you have anything more to do with them. Don't you go buying them sweets, nor balls, nor anything else. You just keep away from that bridge, for a long time. Look, I'll have Mr Guffy deliver what you want in the evenings, if you'd like, just so's you don't have to come across that bridge.'

Mr Fulwood was flabbergasted. 'But why?' he asked.

'Because otherwise they won't leave you alone until they've got what they want,' she said. 'You take care, Mr Fulwood.'

'I'm sure you're exaggerating,' he said, paid for his groceries and left, but not with a bag of sweets. Mrs Guffy simply wouldn't sell one to him.

After a day or so, he began to understand something of what she might have meant. He had been seeing these thoroughly nice boys twice a day, and at the end of that afternoon, as he left, one of them called after him: 'See you later, then, Mr Fulwood.' It was Errol who had called, he thought.

He had not realized that they had meant later that night. He was just watching the television when he heard them outside his back door.

'Mr Fulwood,' he heard a boy's voice call, and recognized it at once as Trevor's.

He went to the kitchen window and opened it. Outside they were all there, all five of them.

'Hello, Trevor. Hello,' he said to the others. 'What do you want? Is something wrong?'

'No, Mr Fulwood, we just wondered if you would like to join us.' That was Errol.

'What, now?'

'Yes.'

'No, I don't think so, not just now.'

But they just stood there, saying nothing.

'No, not just now,' he said again. 'I'll see you tomorrow, no doubt.'

They left, but twice again they came back that night, calling him either at the front door or the back, and the last time was at just after half-past ten, just before he went to bed. He was upstairs in his bedroom at the time, so he opened the window and looked down.

'No, I'm going to bed,' he said firmly, 'where you should be going at this time of night. I'll see you tomorrow. Now go home, or your parents will be worried.'

He didn't like the gentle laughter which arose after that last remark, but closed the window anyway, and they went away. He decided that he had better have a quiet word with them in the morning. This sort of thing could get him a bad name with the neighbours, if the calling had disturbed them. People went to bed early in the village, he knew that.

But the next day they all seemed so happy to have him with them that he had not the heart to tell them off. Instead, he merely mentioned casually that they should not make a noise which might annoy the neighbours at night, and they looked so contrite that he knew he would not have to speak to them about it again.

That afternoon he could not see them, for he had to pay a visit to the town on the bus in order to collect his money from the bank, and the bus did not take him over the bridge but in the opposite direction. When he returned home, he found the vicar waiting on his doorstep. That was a surprise.

'Come in, vicar,' he said, unlocking his door. 'Shall I make you a cup of tea?'

'No, thank you, Mr Fulwood,' said the vicar. For some reason he looked decidedly uncomfortable, and once he had sat down came straight to the point.

'I understand you have been seeing some children on the bridge,' he said, without any preamble.

'Well, everyone knows that,' said Mr Fulwood. 'What's wrong with that? You're the third person to mention them to me, with some kind of warning.'

'Errol, Charlie, Trevor, Mark, Darren—is that right?'

'Well, yes, they are their names. Five of them. But what's wrong?'

'They are evil, Mr Fulwood, that's what is wrong.' The vicar was so serious that Mr Fulwood was startled.

'I just can't believe it. They're so—well, such a joy to be with. They're so lively and well-behaved with it, too.There's no evil there, I assure you.'

'So you intend to go on seeing them, and talking to them, and have them ask you to join them, eh?'

'I don't see why not.'

'When they call tonight, then, I beg you not to open your door to them.'

Now how on earth did the vicar know that they had called at his house? Somebody must have told him. He could not possibly have heard their voices calling him from as far away as the vicarage.

'And then, *not having opened your door to them tonight*, I want you to go over the bridge as usual in the morning, but this time count them.'

'Count them?' Mr Fulwood was astonished. There were five of them, he knew that for certain.

'You might find that tomorrow there will be six. If you find that there are six, come away at once and see me. At once, Mr Fulwood,' the vicar emphasized.

Then he left Mr Fulwood's cottage without further explanation. He simply would not give one. All the vicar would say was that if he would not believe that the boys were evil, if tomorrow there were six of them he might be able to convince him.

That night, they did come again. It was almost as if they were circling the house, and they were calling to him plaintively: 'Mr Fulwood, won't you join us? . . . Mr Fulwood, do come and join us . . . We like your company, Mr Fulwood, so why won't you join us?'

It was somewhat unnerving, and at the same time like ignoring good friends, but nevertheless he remained inside the house with the curtains drawn, not moving until the voices faded and they had gone. They made no attempt to try the door, he noticed. Of course, they were too polite for that, he knew. In the morning he would pretend that he had fallen asleep and had not heard them.

That was his intention as he set out for his regular walk across the bridge and into the main part of the village, but when he reached the bridge they were all seated on the low parapet swinging their legs and with such smiling, welcoming looks on their faces that he thought that he would just walk by. Then he happened to notice that there was an extra boy sitting with them, and almost stopped in his tracks. Mr Fulwood had seen him somewhere around before, but for the moment could not quite place him. Then he noticed several marks scratched deep into the solid stonework of the bridge.

'A lorry ran into it yesterday afternoon while you were out of the village,' Errol explained, seeing where he was gazing.

'Well, I can't stop just now. I'm in a bit of a hurry.'

And he walked on quickly into the village, but instead of going to the shop went straight to the vicarage and rang the doorbell. The vicar answered the door almost at once.

'There were six?' he asked.

'Yes,' said Mr Fulwood. 'Now tell me how you knew that there would be.'

'Come inside,' advised the vicar. Then: 'Describe the sixth boy, would you?'

'Well, he had dark hair, was pleasant-looking—as all of them are—he was wearing a blue jumper, jeans, some sort of sandals on his feet. Oh, and he had a little scar on the tip of his nose. Not much to say about him, really.'

'Tommy Stokes,' snapped the vicar. His tone was almost vicious. 'You are in great danger, Mr Fulwood. I would advise you to leave the village as soon as possible, today if you can, but tomorrow at the latest. Go away—anywhere. Take a holiday, if you like, but stay away for two or three weeks. But whatever you decide to do, don't mention anything about it to any of these boys.'

'Now look here, vicar,' said Mr Fulwood, 'you don't take me for a complete fool, do you? Then perhaps you'll be so kind as to tell me just what is going on!'

The vicar seemed quite calm now. 'Those delightful children

all attended the school here in the village at one time, but they were not delightful at all. It was only discovered what they had been up to when a body was dug up in the middle of the night from the churchyard, and in the morning found draped across the altar of the church, together with other indications that some kind of black magic rite had taken place.

'From then on, things went from bad to worse. Of course, it was discovered who had been involved, but they were held then to be too young to be prosecuted, so nothing was done. Then there was no stopping them. Their last of many activities was to persuade the driver of a van to stop outside the village on some pretext, and give them a lift. On the way, just before the bridge, they attempted to kill the driver. It seems that they hoped to force him to drive straight at the bridge, while at the same time escaping themselves, just in time. Unfortunately for them, the rear doors of the van jammed and they were unable to escape.

'The van crashed, right enough, but the survivor was the driver, not the boys. He was able to tell the story—not all the details, you realize, but enough for everyone to know exactly what had been planned. He was permanently crippled, and now has an artificial leg. You've met him—Mr Cooper.'

'But—' Mr Fulwood was slowly beginning to realize the implication of what the vicar was telling him.

'Yes, Mr Fulwood, I'm afraid so. We know that they are there. We, who have lived in this village for years, know how to avoid them. Your great problem is—you have *seen* them.'

The vicar must have been having him on. It must be some sort of joke! Mr Cooper must be getting on for eighty. What was he doing driving a van?

'It was about thirty years ago. Now do you see? Tommy Stokes also saw them. He was killed in an accident with a lorry yesterday afternoon. He had told his mother only the day before that they had asked him to join them. People who saw the accident said the driver simply seemed to lose all control, and by the time he had recovered, Tommy Stokes was dead under his wheels. The funeral is on Friday. Now do you see, Mr Fulwood? Tommy Stokes was also new to the village, so he didn't know any better, either.

'Pack your things and go at once, Mr Fulwood. Next week, I intend to exorcise that bridge, in hopes that we can at last get rid of them. Come back to your cottage in two or three weeks, and then you should be safe.'

Mr Fulwood almost ran out of the vicarage and back towards his own home. He could hear the vicar calling after him: 'Whatever you do, don't stop on the way!'

He didn't need the advice. By now he was convinced. When he came within sight of the bridge he broke into a gentle trot, then calmed down and determined to walk past as if nothing was wrong.

The children were no longer sitting on the bridge, but standing across the road. They were not exactly in a line as if to prevent his passing, but they were *there*.

'Can't stop now,' he said as cheerily as he could.

The new boy was standing right in front of him.

'Come on, Tommy,' he said, not stopping. 'I'll talk to you later.'

He realized his mistake at once.

'He knows!' shouted Errol.

Now they did form a line like a barricade, to prevent him from moving further. He began to push through them, in a panic. He got through, then began to run. The boys began to run after him. To the house, he was thinking. They don't come into the house. Once there I'll be safe. He ran faster, and looked behind to see if they were still chasing, but they were not any longer. In fact, they were no longer there at all.

The car coming round the bend in the road hit him almost before he saw it, and then they were all round him again, looking at him as he lay on the ground. Charlie helped him to his feet, and Trevor and Mark took hold of his hands to help steady him. They were never evil children, he was convinced of that now. That vicar ought to be in a lunatic asylum. He had heard of vicars like that, who went mad.

The driver of the car was climbing, rather shaken, out of his vehicle and coming towards him. There seemed to be something lying on the ground in front of the car, and the driver stopped and bent down over it. Then Mr Fulwood realized that it was his own body lying there. He shook himself free of the helping hands and stared at the boys wildly.

'You were too late, Mr Fulwood,' said Errol calmly, with a most attractive smile on his face. 'Now you've joined us properly.'

Rats

M. R. JAMES

*'And if you was to walk through the bedrooms now, you'd see the
ragged, mouldy bedclothes a-heaving and a-heaving like seas.'
'And a-heaving and a-heaving with what?' he says.
'Why, with the rats under 'em.'*

But was it with the rats? I ask, because in another case it
was not. I cannot put a date to the story, but I was young
when I heard it, and the teller was old. It is an ill-
proportioned tale, but that is my fault, not his.

It happened in Suffolk, near the coast. In a place where the
road makes a sudden dip and then a sudden rise; as you go
northward, at the top of that rise, stands a house on the left of
the road.

It is a tall red-brick house, narrow for its height; perhaps it
was built about 1770. The top of the front has a low triangular
pediment with a round window in the centre. Behind it are
stables and offices, and such garden as it has is behind them.
Scraggy Scotch firs are near it: an expanse of gorse-covered land
stretches away from it. It commands a view of the distant sea
from the upper windows of the front. A sign on a post stands
before the door; or did so stand, for though it was an inn of
repute once, I believe it is so no longer.

To this inn came my acquaintance, Mr Thomson, when he
was a young man, on a fine spring day, coming from the
University of Cambridge, and desirous of solitude in tolerable
quarters and time for reading. These he found, for the landlord
and his wife had been in service and could make a visitor
comfortable, and there was no one else staying in the inn. He
had a large room on the first floor commanding the road and the
view, and if it faced east, why, that could not be helped; the
house was well built and warm.

He spent very tranquil and uneventful days: work all the
morning, an afternoon perambulation of the country round, a

little conversation with country company or the people of the inn in the evening over the then fashionable drink of brandy and water, a little more reading and writing, and bed; and he would have been content that this should continue for the full month he had at disposal, so well was his work progressing, and so fine was the April of that year—which I have reason to believe was that which Orlando Whistlecraft chronicles in his weather record as the 'Charming Year'.

One of his walks took him along the northern road, which stands high and traverses a wide common, called a heath. On the bright afternoon when he first chose this direction his eye caught a white object some hundreds of yards to the left of the road, and he felt it necessary to make sure what this might be. It was not long before he was standing by it, and found himself looking at a square block of white stone fashioned somewhat like the base of a pillar, with a square hole in the upper surface. Just such another you may see at this day on Thetford Heath.

After taking stock of it he contemplated for a few minutes the view, which offered a church tower or two, some red roofs of cottages and windows winking in the sun, and the expanse of sea—also with an occasional wink and gleam upon it—and so pursued his way.

In the desultory evening talk in the bar, he asked why the white stone was there on the common.

'An old-fashioned thing, that is,' said the landlord (Mr Betts), 'we was none of us alive when that was put there.'

'That's right,' said another.

'It stands pretty high,' said Mr Thomson. 'I dare say a sea-mark was on it some time back.'

'Ah! yes,' Mr Betts agreed, 'I 'ave 'eard they could see it from the boats; but whatever there was, it's fell to bits this long time.'

'Good job too,' said a third, ''twarn't a lucky mark, by what the old men used to say; not lucky for the fishin', I mean to say.'

'Why ever not?' said Thomson.

'Well, I never see it myself,' was the answer, 'but they 'ad some funny ideas, what I mean, peculiar, them old chaps, and I shouldn't wonder but what they made away with it theirselves.'

It was impossible to get anything clearer than this: the company, never very voluble, fell silent, and when next someone spoke it was of village affairs and crops. Mr Betts was the speaker.

Not every day did Thomson consult his health by taking a country walk. One very fine afternoon found him busily writing at three o'clock. Then he stretched himself and rose, and walked out of his room into the passage.

Facing him was another room, then the stair-head, then two more rooms, one looking out to the back, the other to the south. At the south end of the passage was a window, to which he went, considering with himself that it was rather a shame to waste such a fine afternoon. However, work was paramount just at the moment; he thought he would just take five minutes off and go back to it; and those five minutes he would employ—the Bettses could not possibly object—to looking at the other rooms in the passage, which he had never seen.

Nobody at all, it seemed, was indoors; probably, as it was market day, they were all gone to the town, except perhaps a maid in the bar. Very still the house was, and the sun shone really hot; early flies buzzed in the window-panes. So he explored.

The room facing his own was undistinguished except for an old print of Bury St Edmunds; the two next him on his side of the passage were gay and clean, with one window apiece, whereas his had two. Remained the south-west room, opposite to the last which he had entered. This was locked; but Thomson was in a mood of quite indefensible curiosity, and feeling confident that there could be no damaging secrets in a place so easily got at, he proceeded to fetch the key of his own room, and when that did not answer, to collect the keys of the other three. One of them fitted, and he opened the door.

The room had two windows looking south and west, so it was as bright and the sun as hot upon it as could be. Here there was no carpet, but bare boards; no pictures, no washing-stand, only a bed, in the farther corner: an iron bed, with mattress and bolster, covered with a bluish check counterpane. As featureless a room as you can well imagine, and yet there was something that made Thomson close the door very quickly and yet quietly behind him and lean against the window-sill in the passage, actually quivering all over.

It was this, that under the counterpane someone lay, and not only lay, but stirred. That it was some *one* and not some *thing* was certain, because the shape of a head was unmistakable on the bolster; and yet it was all covered, and no one lies with

covered head but a dead person; and this was not dead, not truly dead, for it heaved and shivered.

If he had seen these things in dusk or by the light of a flickering candle, Thomson could have comforted himself and talked of fancy. On this bright day that was impossible. What was to be done? First, lock the door at all costs. Very gingerly he approached it and bending down listened, holding his breath; perhaps there might be a sound of heavy breathing, and a prosaic explanation. There was absolute silence. But as, with a rather tremulous hand, he put the key into its hole and turned it, it rattled, and on the instant a stumbling padding tread was heard coming towards the door.

Thomson fled like a rabbit to his room and locked himself in: futile enough, he knew it was; would doors and locks be any obstacle to what he suspected? But it was all he could think of at the moment, and in fact nothing happened; only there was a time of acute suspense—followed by a misery of doubt as to what to do.

The impulse, of course, was to slip away as soon as possible from a house which contained such an inmate. But only the day before he had said he should be staying for at least a week more, and how if he changed plans could he avoid the suspicion of having pried into places where he certainly had no business? Moreover, either the Bettses knew all about the inmate, and yet did not leave the house, or knew nothing, which equally meant that there was nothing to be afraid of, or knew just enough to make them shut up the room, but not enough to weigh on their spirits: in any of these cases it seemed that not much was to be feared, and certainly so far he had had no sort of ugly experience. On the whole the line of least resistance was to stay.

Well, he stayed out his week. Nothing took him past that door, and, often as he would pause in a quiet hour of day or night in the passage and listen, and listen, no sound whatever issued from that direction. You might have thought that Thomson would have made some attempt at ferreting out stories connected with the inn—hardly perhaps from Betts, but from the parson of the parish, or old people in the village; but no, the reticence which commonly falls on people who have had strange experiences, and believe in them, was upon him.

Nevertheless, as the end of his stay drew near, his yearning after some kind of explanation grew more and more acute. On his solitary walks he persisted in planning out some way, the least obtrusive, of getting another daylight glimpse into that room, and eventually arrived at this scheme. He would leave by an afternoon train—about four o'clock. When his fly was waiting, and his luggage on it, he would make one last expedition upstairs to look round his own room and see if anything was left unpacked, and then, with that key, which he had contrived to oil (as if that made any difference!), the door should once more be opened, for a moment, and shut.

So it worked out. The bill was paid, the consequent small talk gone through while the fly was loaded: 'pleasant part of the country—been very comfortable, thanks to you and Mrs Betts—hope to come back some time,' on one side: on the other, 'very glad you've found satisfaction, sir, done our best—always glad to 'ave your good word—very much favoured we've been with the weather, to be sure.' Then, 'I'll just take a look upstairs in case I've left a book or something out—no, don't trouble, I'll be back in a minute.' And as noiselessly as possible he stole to the door and opened it.

The shattering of the illusion! He almost laughed aloud. Propped, or you might say sitting, on the edge of the bed was—nothing in the round world but a scarecrow! A scarecrow out of the garden, of course, dumped into the deserted room . . . Yes; but here the amusement ceased. Have scarecrows bare bony feet? Do their heads loll on to their shoulders? Have they iron collars and links of chain about their necks? Can they get up and move, if never so stiffly, across a floor, with wagging head and arms close at their sides? and shiver?

The slam of the door, the dash to the stair-head, the leap downstairs, were followed by a faint. Awaking, Thomson saw Betts standing over him with the brandy bottle and a very reproachful face. 'You shouldn't a done so, sir, really you shouldn't. It ain't a kind way to act by persons as done the best they could for you.'

Thomson heard words of this kind, but what he said in reply he did not know. Mr Betts, and perhaps even more Mrs Betts, found it hard to accept his apologies and his assurances that he would say no word that could damage the good name of the house. However, they *were* accepted.

Since the train could not now be caught, it was arranged that

Thomson should be driven to the town to sleep there. Before he went the Bettses told him what little they knew.

'They says he was landlord 'ere a long time back, and was in with the 'ighwaymen that 'ad their beat about the 'eath. That's how he come by his end: 'ung in chains, they say, up where you see that stone what the gallus stood in. Yes, the fishermen made away with that, I believe, because they see it out at sea and it kep' the fish off, according to their idea. Yes, we 'ad the account from the people that 'ad the 'ouse before we come. "You keep that room shut up," they says, "but don't move the bed out, and you'll find there won't be no trouble."

'And no more there 'as been; not once he haven't come out into the 'ouse, though what he may do now there ain't no sayin'. Anyway, you're the first I know on that's seen him since we've been 'ere: I never set eyes on him myself, nor don't want. And ever since we've made the servants' rooms in the stablin', we ain't 'ad no difficulty that way. Only I do 'ope, sir, as you'll keep a close tongue, considerin' 'ow an 'ouse do get talked about': with more to this effect.

The promise of silence was kept for many years. The occasion of my hearing the story at last was this: that when Mr Thomson came to stay with my father it fell to me to show him to his room, and instead of letting me open the door for him, he stepped forward and threw it open himself, and then for some moments stood in the doorway holding up his candle and looking narrowly into the interior. Then he seemed to recollect himself and said: 'I beg your pardon. Very absurd, but I can't help doing that, for a particular reason.' What that reason was I heard some days afterwards, and you have heard now.

The Haunted Mill

JEROME K. JEROME

Well, you all know my brother-in-law, Mr Parkins (began Mr Coombes, taking the long clay pipe from his mouth, and putting it behind his ear; we did not know his brother-in-law, but we said we did, so as to save time), and you know of course that he once took a lease of an old mill in Surrey, and went to live there.

Now you must know that, years ago, this very mill had been occupied by a wicked old miser, who died there, leaving—so it was rumoured—all his money hidden somewhere about the place. Naturally enough, everyone who had since come to live at the mill had tried to find the treasure; but none had ever succeeded, and the local wiseacres said that nobody ever would, unless the ghost of the miserly miller should, one day, take a fancy to one of the tenants, and disclose to him the secret of the hiding-place.

My brother-in-law did not attach much importance to the story, regarding it as an old woman's tale, and, unlike his predecessors, made no attempt whatever to discover the hidden gold.

'Unless business was very different then from what it is now,' said my brother-in-law, 'I don't see how a miller could very well have saved anything, however much a miser he might have been: at all events, not enough to make it worth the trouble of looking for it.'

Still, he could not altogether get rid of the idea of that treasure.

One night he went to bed. There was nothing very extraordinary about that, I admit. He often did go to bed of a night. What *was* remarkable, however, was that exactly as the clock of the village church chimed the last stroke of twelve, my brother-in-law woke up with a start, and felt himself quite unable to go to sleep again.

Joe (his Christian name was Joe) sat up in bed, and looked around.

At the foot of the bed something stood very still, wrapped in shadow.

It moved into the moonlight, and then my brother-in-law saw that it was a figure of a wizened little old man, in knee-breeches and a pig-tail.

In an instant the story of the hidden treasure and the old miser flashed across his mind.

'He's come to show me where it's hid,' thought my brother-in-law; and he resolved that he would not spend all this money on himself, but would devote a small percentage of it towards doing good to others.

The apparition moved towards the door: my brother-in-law put on his trousers and followed it. The ghost went downstairs into the kitchen, glided over and stood in front of the hearth, sighed and disappeared.

Next morning, Joe had a couple of bricklayers in, and made them haul out the stove and pull down the chimney, while he stood behind with a potato-sack in which to put the gold.

They knocked down half the wall, and never found so much as a four-penny bit. My brother-in-law did not know what to think.

The next night the old man appeared again, and again led the way into the kitchen. This time, however, instead of going to the fireplace, it stood more in the middle of the room, and sighed there.

'Oh, I see what he means now,' said my brother-in-law to himself; 'it's under the floor. Why did the old idiot go and stand up against the stove, so as to make me think it was up the chimney?'

They spent the next day in taking up the kitchen floor; but the only thing they found was a three-pronged fork, and the handle of that was broken.

On the third night, the ghost reappeared, quite unabashed, and for a third time made for the kitchen. Arrived there, it looked up at the ceiling and vanished.

'Umph! he don't seem to have learned much sense where he's been to,' muttered Joe, as he trotted back to bed; 'I should have thought he might have done that first.'

Still, there seemed no doubt now where the treasure lay, and the first thing after breakfast they started pulling down the ceiling. They got every inch of the ceiling down, and they took up the boards of the room above.

They discovered about as much treasure as you would expect to find in an empty quart-pot.

On the fourth night, when the ghost appeared, as usual, my brother-in-law was so wild that he threw his boots at it; and the boots passed through the body, and broke a looking-glass.

On the fifth night, when Joe awoke, as he always did now at twelve, the ghost was standing in a dejected attitude, looking very miserable. There was an appealing look in its large sad eyes that quite touched my brother-in-law.

'After all,' he thought, 'perhaps the silly chap's doing his best. Maybe he has forgotten where he really did put it, and is trying to remember. I'll give him another chance.'

The ghost appeared grateful and delighted at seeing Joe prepare to follow him, and led the way into the attic, pointed to the ceiling, and vanished.

'Well, he's hit it this time, I do hope,' said my brother-in-law; and next day they set to work to take the roof off the place.

It took them three days to get the roof thoroughly off, and all they found was a bird's nest; after securing which they covered up the house with tarpaulins, to keep it dry.

You might have thought that would have cured the poor fellow of looking for treasure. But it didn't.

He said there must be something in it all, or the ghost would never keep coming as it did; and that, having gone so far, he would go on to the end, and solve the mystery, cost what it might.

Night after night, he would get out of his bed and follow that spectral old fraud about the house. Each night, the old man would indicate a different place; and, on each following day, my brother-in-law would proceed to break up the mill at the point indicated, and look for the treasure. At the end of three weeks, there was not a room in the mill fit to live in. Every wall had been pulled down, every floor had been taken up, every ceiling had had a hole knocked in it. And then, as suddenly as they had begun, the ghost's visits ceased; and my brother-in-law was left in peace, to rebuild the place at his leisure.

'What induced the old image to play such a silly trick upon a family man and a ratepayer? Ah! That's just what I cannot tell you.'

Some said that the ghost of the wicked old man had done it to punish my brother-in-law for not believing in him at first; while others held that the apparition was probably that of some deceased local plumber and glazier, who would naturally take an interest in seeing a house knocked about and spoilt. But nobody knew anything for certain.

The Last Bus

J. M. JOHNSON-SMITH

He left the friendly warmth of the pub, voices following him—a temporary link with civilization—then the door clicked shut and he was alone.

His footsteps echoed crisply in the cold night air, his frosty breath adding to the low mist that swirled around his legs. He walked purposefully, though uncertain of his journey across the moor. Helpful directions from his drinking companions echoed confusedly in his brain. He was alone—in the middle of nowhere. He walked automatically, keeping to the centre of the road, hands thrust deep into his pockets.

A persistent throbbing penetrated his thoughts. He stopped. His last footstep echoed and was gone. Through the gloom, he saw low headlights approaching. The familiar shape of a single-decker bus glided through the mist and stopped by his side. He boarded it swiftly.

'Good evening.' His greeting hung in the air, unwanted by the conductor and driver. The bus roared into life and sped on its way as he gingerly made his way to a seat. Inside was dimly lit, and a strange smell pervaded his nostrils.

He looked up as the conductor approached.

'Bolters Farm, please.' He produced a ten pence piece from his pocket. 'This is all I've got, is it enough?'

The conductor remained silent, his face hidden in the shadow of his peaked cap. He took a ticket from his clip and punched it, and held it out. His hands glowed red in the dim light. Blood red. Blood trickled slowly down from inside a tattered sleeve.

For a moment there was a shocked silence.

'Hey, your arm's bleeding . . .' he began.

There was no reply. The conductor walked back to the driver, leaving him clutching nervously at his ticket.

The bus creaked and jolted alarmingly as it rushed through the night. He looked down at the ticket grasped in his cold fingers. It looked unfamiliar. He held it up to the dull light above his seat. A strange uneasiness filled him as he gazed at the old-fashioned design which was so long out of date. In a sudden panic, he thrust the offending ticket deep into his pocket and peered out of the window.

Friendly lights streamed out of the unshuttered windows of a farmhouse. There was one other passenger on the bus. He had noticed him huddled in the seat behind. He turned quickly to ask if the house was Bolters Farm.

The man raised his head and leered at him with sightless eyes. Again the musty smell surrounded him—a smell of death.

He stumbled to his feet, his half-paralysed legs dragging along the rutted bus floor.

'Driver,' he cried, his voice raised in terror. 'Driver, stop, I want to get off.'

The bus roared on. The driver turned his head. He saw eyeless sockets and pulp beneath the peaked cap. The conductor laughed silently, mirthlessly, as, in terror of his life, he flung himself from the bus.

He landed on a grass bank, and lay there, gibbering incoherently. Time passed, minutes or hours. He re-lived the nightmare over and over again as he lay there in the cold mist.

A sudden light shone in his eyes. He protected them with a shaking arm.

'Leave me,' he pleaded hysterically, 'I've done you no harm.'

'Now then, old chap,' the voice was warm and comforting, and he felt strong arms around his shaking body.

'You've been out with the boys,' the voice chided him gently, as expert fingers slid over him. 'There, no bones broken.'

He forced himself to look at the owner of the voice. A kindly gaze met his terror-stricken eyes.

'You weren't on the bus?' he queried urgently.

'Bus? What bus?'

He passed a hand across his face. 'That bus . . .' he lapsed into silence and began to tremble violently.

'Steady on,' the voice spoke authoritatively. 'I'm a doctor, I live just down the road. Now, how did you get here?'

He began to explain, quietly at first, until the memory crowded in on him.

'The bus stopped by me—so I got on—oh, sweet Jesus, it was horrible!' he groaned.

'What type of bus was it?' The doctor's tone of voice had changed. Was he frightened, too?

'A single-decker, badly lit inside—that's why I didn't notice their faces.'

'Faces?' the doctor queried sharply.

'Faces!' he almost screamed the word. 'Bloody, smashed-up faces.'

The doctor paled. 'Come on,' he said. 'You've been drinking too much.'

Something in the doctor's eyes gave him away.

'What is it? Tell me,' he clutched the doctor's coat, demanding an explanation.

'There aren't any buses on this route,' the doctor replied flatly, 'the last one crashed many years ago, and the service was discontinued.'

'It crashed?' His voice rose. 'Who was on it?'

'Two crew and a passenger. They were all killed.'

The musty smell—that was it. An old bus, filled with the stench of death. His stomach turned convulsively. He wanted to retch. Three bloody, pulpy faces swam before his eyes.

A strong hand pushed his head between his shaking knees.

'You'll be all right in a minute.' The doctor's voice had returned to its normal, reassuring tone. The sickness passed. He felt the chill air eating into his bones as sweat lay cold on his skin.

'Better?'

He nodded. The doctor sighed with apparent relief.

'I'll take you home,' he said. 'You've had a fine old nightmare by the sound of it while you slept the drinks off.'

'But the bus . . .' he began.

'No bus,' the doctor laughed, 'just a bad dream.'

A dream? Perhaps the doctor was right. What a fool he was to have been so upset. It had all seemed so real. The bus had stopped, and he had boarded it, and asked for a ticket, and paid a—

'That's it,' he shouted, 'I gave the conductor a ten pence piece for my fare. It was all I had.'

The doctor watched him closely as he searched feverishly through his pockets. His fingers explored every corner. He felt light-headed as the hard surface of a coin met his exploring fingers. He laughed shakily, withdrawing the coin from his pocket.

'There,' he said, shamefaced. 'You were right, it *was* a nightmare.'

For a moment he thought he detected a hint of relief in the doctor's laugh as it rang out in the quiet night.

'Let's go.' The doctor helped him to his feet. His legs were still unsteady.

He returned the coin to his pocket. His spine froze as he encountered paper. He pulled it slowly from his pocket. In the lights from the doctor's car he could see it was an old-fashioned bus ticket.

The blood on it was still wet.

The Scene

of the

Crime

GERALD KERSH

The Big Man with the Little Black Bag turned to the right at The Bricklayers, walked on and turned right again at St George's Church. Then he found that he was lost. Someone had misdirected him, or he had misinterpreted the direction. He stood in a small curved street of cheap and pretentious houses with plaster columns, and basements fenced in with massive, spear-headed iron railings. The snow, trampled to slush in the main road, lay here like a sheepskin rug. Something like a pancake of yellow light lay under every lamp-post. The Big Man with the Little Black Bag was aware of a certain uneasiness.

Then he saw the policeman.

The policeman was standing in the penumbra beyond one of the circles of light. The top of his helmet bobbed seven feet above the snow. He was enormous in his heavy greatcoat. The Big Man with the Little Black Bag approached, with a certain trepidation, and said: 'I beg pardon, but I seem to have lost my way. Can you tell me the way to Mahogany Road?'

The policeman replied: 'Mahogany Road, sir? Yes, sir. Let me see now, Mahogany Road. This is Tulip Crescent. Follow the Crescent around to your left, just as you're going, then bear right along Jade Street, and when you get to a public-house called The Jolly Farmers, turn sharp right and there you are.'

The Big Man with the Little Black Bag said: 'Did you say Tulip Crescent? Now where have I heard Tulip Crescent mentioned before?'

The policeman said: 'I daresay you would have heard of it in connection with the Joyce Murder.'

'Oh yes, yes, the Joyce Murder, Tulip Crescent. Of course,' said the other uneasily. 'Somewhat before my time, I believe.'

'I daresay it would be,' said the policeman, 'but I remember it.'

'It happened at No. 14, I believe?'

'Yes, sir, it happened at No. 14. But after the scandal they changed the number of the house, and the number is now 13b. Yes, sir, if you want to get to Mahogany Road, follow the Crescent around to your left, then bear right along Jade Street, and when you get to a public-house called The Jolly Farmers, turn sharp right and there you are.'

The Big Man with the Little Black Bag walked on, and the policeman walked with him. From time to time, the policeman flashed his lamp into a doorway. Once he tried a lock.

'Live near here?' he asked.

The other replied: 'No, I live in Australia, in Sydney—near Sydney, at least. Not far from Sydney. I daresay you wonder what brings me here. Well, as a matter of fact, it *is* rather extraordinary but I have to go to Mahogany Road to meet a distant relation of mine who I have never met before. It's a family affair. And so *this* is Tulip Crescent?'

The policeman said, with something like relish: 'This is Tulip Crescent.'

He turned and threw the beam of his light upon the door of a house.

The Big Man with the Little Black Bag saw 13b in brass against green paint in a halo of dried-up metal polish. 'It was before my time,' he said.

The policeman eased his knees and beat his gloved hands together, and said: 'It was quite a case, sir, as you may have read.'

'An old lady and an old gentleman were murdered, I believe, and no one ever found out who did it,' said the stranger.

'That's right, and after thirty years the murderer is still undiscovered,' said the policeman.

'It means that he is still at large,' said the Big Man with the Little Black Bag. 'Still at large.'

The policeman cleared his throat judicially and said: 'At least he has never been brought to court. You understand, sir, that this is my beat and I know a good deal about it. It is my business to know what goes on, and what has gone on, sir, on my beat.'

'But this was thirty years ago.'

'Yes, sir, but we have our proper professional interest in these things. I am only an ordinary police constable, as you see. But these things are interesting. And if, as I might say, I walk up and down past No. 14—or, I should say, 13b—many times every night, it is only natural for me to take an interest. You know the facts of the case, I presume?'

The Big Man with the Little Black Bag, shifting from foot to foot and looking nervously up and down the deserted Crescent, said: 'An old lady and her brother were murdered for their money, and the murderer got away scot-free, that's all I know. And now if you'll excuse me—'

The policeman, beating away the cold with his gloved hands, said: 'It was an interesting case, sir. Mr Spoon of the *Sunday Special* wrote very intelligently about it in his *Unsolved Murder Mysteries*. There was Miss Joyce and her brother, Mr Joyce. They lived in what was then No. 14, you see, sir. The Joyces were people of independent means. But I daresay you've read all about this in *Unsolved Murder Mysteries*, by the gentleman whose name I have already mentioned.'

'No, no.'

'The father was an actuary, whatever that may be, sir.'

'I believe it is a man who guesses the odds in insurance, officer.'

'I see you're a betting man, sir,' said the policeman.

'No, no, not at all,' said the Big Man with the Little Black Bag. 'I never bet.'

The policeman continued: 'Their father was an actuary, and he left the lady, Miss Joyce, a considerable sum of money. Her brother, Mr Joyce, was, if I may say so, sir, a briefless barrister, who liked his bottle and got around to living on his sister—if you get what I mean, sir.'

'Yes, yes.'

'You understand, of course, that I have studied all this pretty closely, sir, this being my beat.'

'Naturally.'

'The old lady—she was a few years older than her brother— didn't trust banks and kept most of her money, or at least a good deal of it, in the house. A bad principle,' said the policeman. 'For my part, give me a bank.'

'Give me a bank,' said the Big Man with the Little Black Bag.

'Well, sir, the inevitable, as they say in the newspapers, is bound to happen. One dark night there was a murder,' said the policeman with gusto, 'a murder of almost unprecedented brutality. The criminal went into No. 14 as it was then, murdered the old lady, murdered the old gentleman and ran off with every farthing he could lay his hands on. It amounted to something like six hundred and twenty-five pounds six shillings and threepence.'

The other said: 'Where was the policeman on duty here, at that time?'

The policeman replied: 'This was over thirty years ago, sir, and things were a little looser in the Force then than they are at the present time. True, your police constable had to account for his movements, but there was not quite the check-up then that you get now. Your sergeants came round of course. But the policeman who was on duty here at the time of the Joyce Murder had the whole Crescent to cover and this Crescent makes a considerable bend, as you'll see by the time you get to Mahogany Road. The man who killed poor Miss Joyce and her poor brother must have watched the movements of the policeman on the beat. You know that we are usually at a certain place within a certain time. The whole thing was done in a few minutes. A few minutes? One minute, two minutes at the most. The culprit had means of entry to the house, went in, did his work, and came out, shutting the door behind him. We should not have discovered the fact for days, perhaps, if the old gentleman—I mean Miss Joyce's brother—had not managed to drag himself to the front door before he died. How he managed to do it nobody knows. His skull was smashed. It was a horrible affair, as you may well imagine, and that is why they changed the number of the house. Then, of course, there was no housing shortage, and people preferred not to live in a place where such things had happened. And so, as the gentleman said in the newspaper, it remains an unsolved mystery.'

The Big Man with the Little Black Bag said: 'Were there no clues?'

The policeman replied: 'Not one. It happened about ten o'clock one night in September. If it had been wet, there might have been footprints. It was one of those dry, dusty nights, with a bit of a breeze blowing. The officer on duty observed that this

crime had been committed because, in the first place, he saw that the door of No. 14 was shut. In the second place . . .'

'Did you say *shut*?' asked the Big Man with the Little Black Bag.

'Yes, sir. He was accustomed to hear from Mr and Miss Joyce at about that time. The old gentleman was not afraid of anything. He was a lawyer. But the old lady, who kept the purse-strings, she liked to be friendly with the police and, therefore, she made a point of giving the officer on duty a cake, or a pie or a sandwich—sometimes a glass of rum—late in the evening. For the first time in seven years there was silence in No. 14. The officer on duty stopped at the door of No. 14, flashing his lamp about and wondering what had happened. Then, all of a sudden, he saw a red blob creeping out under the front door. He knocked, and nobody answered, he rammed his shoulder against the door, nearly fell inside because the door was not quite caught. Then he saw poor old Mr Joyce with his head beaten in on the doormat.'

The stranger said: 'And there was an end of the matter?'

The policeman nodded. 'There was the end of the matter. Nobody found anything. There was no evidence. They questioned Miss Joyce's nephew, but he was paralysed and had been in bed for the past three years, so *he* was out of the question. It was one of those cases in which Scotland Yard was baffled, sir, baffled. The man was never caught.'

The Big Man with the Little Black Bag glanced uneasily from left to right and said: 'Do you believe that criminals—I mean murderers—feel, as they say, compelled to come back to the scene of the crime?'

'Some do, and some do not,' said the policeman.

The stranger said: 'I can understand that a man might want to come back. I can't imagine what for, but I can imagine a sort of nervous compulsion. I can simply *imagine* it, mind you...'

The policeman said: 'You're living in Mahogany Road?'

The stranger said: 'Which reminds me that I must be getting along. Oh, by the way, what happened to the officer on duty at that time?'

The policeman said: 'He was dismissed the Force. He used to go into the doorway of No. 102 to have a smoke. On the night of the murder it seems he had more than one smoke. He failed in his duty, sir.'

'So the criminal was never found.'

'Never found, no. But what you were saying about murderers coming back, sir. It really does happen sometimes.'

'I daresay,' said the Big Man with the Little Black Bag. 'Were there no finger-prints?'

'The murderer was wearing—'

The policeman looked down at his gloved hands.

'The policeman could have done it,' said the stranger.

'He did,' said the policeman.

Then the Big Man with the Little Black Bag found himself alone in the Crescent, and there were no footprints in the snow except his own.

'Tain't So

MARIA LEACH

Old Mr Dinkins was very ill, so they sent for the doctor. When the doctor came, old man Dinkins said, 'There's nothing the matter with me!'

'You are dying,' said the doctor.

''Tain't so!' said old man Dinkins. But the next day he was dead.

So they put the old man in his coffin; they carried him to church and had his funeral; then they carried him to the graveyard and buried him.

The next morning a neighbour passing the graveyard on his way to work saw old man Dinkins sitting on the graveyard fence.

'Hello, there! I thought you were dead,' said the neighbour.

''Tain't so!' said old man Dinkins.

The neighbour went and told old Mrs Dinkins that her husband was sitting on the graveyard fence and said he was *not* dead.

'Pay no attention,' said the widow. 'He's foolish.'

Later on another neighbour passing by the graveyard heard someone say, 'Hello, Tom!'

'Hello,' said Tom and stopped for a chat. 'It's you, is it?'

'Sure,' said old man Dinkins.

'I heard you were dead.'

''Tain't so!'

'I heard about the burial.'

'Well, you can see I'm not buried.'

'That's so,' said the neighbour and went on his way, somewhat puzzled.

The next day one of the townsmen was passing by the graveyard on horseback. He heard someone say, 'Hello,' and stopped to see who it was. He saw a very old gentleman sitting on the fence, who said, 'What's the news from town?'

'Not much news, except old man Dinkins is dead.'

''Tain't so!'

'That's what they said.'

'Well, 'tain't so.'

'How do you know?' said the man.

'I'm Dinkins.'

'Oh!' said the man and rode away from the place pretty fast.

He stopped at the next store and said, 'There's a funny old fellow sitting on the graveyard fence who says he is old man Dinkins.'

'Can't possibly be,' said the storekeeper.

'Why not?'

'Because old man Dinkins is dead.'

This kept going on week after week, month after month. The whole town knew that old man Dinkins was dead; but old man Dinkins sat on the graveyard fence saying, ''Tain't so.'

After much talk and consultation the townspeople decided to hold *another* burial service.

So they said the burial service over the old man's grave for a second time and set up his gravestone.

The words on the gravestone said:

Here Lies the Body of
Theodore Dinkins
aged 91
Respected citizen of
Wadmalaw Island
who died
January 17, 1853

The next day when old man Dinkins crawled out of his grave, he read what the stone said. He read it over two or three times.

'Well—maybe so,' he said. He hasn't yelled at anybody from the graveyard fence since then.

The Servant

MICHAEL MACLIAMMOIR

It was in London in the year 1916: a month or so before the Easter rising. A very young boy, yet I was an experienced professional: I had been on the stage for nearly six years! And among my many grown-up friends—stage children are often more at home with adults than with each other—was an English actor I will call Kenneth Dane, though only the first name was his own.

A good actor, a French scholar, an ardent convert to the Catholic Church, an enthusiast about all things Irish, and a friend of Mabel Beardsley, the sister of the great artist. Altogether an interesting, companionable and charming man, who, although he was scarcely thirty, seemed to me almost middle-aged and an authority on all things under the sun.

On account, I suppose, of his Catholic faith, he was given a commission in an Irish Regiment of the British Army—we may as well call it the Munster Fusiliers—though he had never set a foot in Ireland in his life and on this chilly weekend of early spring in 1916 (he was in England for a few days' leave) he and I met at a weekend party in a lovely Georgian house on the river near Richmond.

The party was gay, noisy and sentimental with an atmosphere of 'let us be merry for tomorrow we die' after the manner of the times. Kenneth and I shared the bedroom lit by candles and lamps: electricity had not yet penetrated to the suburbs, and on the third night of our stay, as we were preparing for sleep, he said to me: 'Do you ever get premonitions? I do. I know, for example, that I'll never see you—any of you—on this earth again, because I'm going to be killed, you see. Oh, yes, I'm quite certain of it. In fact, I could say "I know".'

My head at this period of my life was much more filled with ghost stories than with sentiments of friendship and my thoughts expressed themselves—as is the habit with my thoughts—immediately. 'Promise me,' I said, 'that if you do get killed you will come and tell me so.'

Suddenly he looked at me over the candlelight that separated our two beds. I was smoking my first cigarette and I was immensely proud of it. He stubbed his out on an ashtray. 'I'll come if God allows me—I'll come and tell you.'

I slept, youthful, heartless, and untroubled. When I woke next morning the bed next to mine was empty: his books, his clothes, his shaving things: all had been taken away, there was a faint smell of brilliantine in the air, that was all. And less than a week later, we received the news! He had been killed three days after his return to France.

Here I must repeat: I was youthful, heartless, and untroubled. No ghost had come to me: no message from the dead had arrived, no premonition had touched me, there had been no sign. And, as the years went by, in spite of our youthful vows of eternal friendship, I confess I forgot all about him.

It was twenty years later—in 1936—that Hilton Edwards and I, who some eight years previously had established the Dublin Gate Theatre and done most of our early work, were sharing a flat in a certain street that had a church like a pepper-pot at the end of it. It was early spring weather, dark, gloomy, and ominous: we had engaged a new man-servant, unknown to us but for his references from Trinity College and an admission— or was it boast?—that he had once served in the British Army. He was interviewed and I felt he would be satisfactory.

'Don't knock on the door when you come up in the morning at nine,' I told him. 'Just bring in the coffee, open the curtains, and that will wake me.'

'Yes, sir.'

Next morning, although after our heavy work in the theatre I sleep well as a rule, I awoke long before nine. I was vaguely disturbed: I wondered why, and finally concluded that it was because of the new man. Would he burn our trousers as he pressed them as his predecessor had done? Would he fill the bath to overflowing, make the coffee all wrong? Presently I heard him coming up the stairs, a church clock somewhere was striking the hour of nine, I remember thinking: 'He's punctual, anyway.'

The door opened softly and a British officer entered the room carrying a coffee tray.

I stared. The room was still dim with its drawn curtains and faint wintry light outside, but I could see him plainly. I thought: 'What is a British officer doing in Dublin in 1936?'

The man advanced into the room, and I saw that it was Kenneth Dane. In spite of the twenty years that had passed by since he died, there was no mistaking the spare figure: the light brown, smoothly-brushed hair: the lean, clean-shaven face. I almost shouted in my astonishment: 'Kenneth! Kenneth Dane!' And it was only then that I remembered his promise to me when I was a boy, and I said: 'You have come back to see me—after all!'

He nodded his head gravely: his lips, half smiling, moved as if in speech, but I heard no sound. He came close to the bed, set down the coffee tray on the table, and still with that old smile on his moving lips, still watching me intently, he turned and opened the curtains.

As the grey morning light streamed into the room, then that khaki-clad figure of the officer seemed to melt rapidly into that of a middle-aged, slightly stooping man: grey haired, and dressed in black trousers and a white linen coat. It was the new servant.

That was the only moment, I think, in which I felt any dismay. I gasped and almost wept, full of a strange, remote anguish: but, seeing the man staring at me in amazement, I made an effort, and muttered some foolish thing about a dream. But the servant's face was white.

'Excuse me, sir: but did you not call me Kenneth Dane, just now when I was coming in?'

'I did, yes . . . I'd been dreaming about a very old friend of mine . . .'

'Excuse me, sir: but it wouldn't be Captain Kenneth Dane, sir, would it?'

'Yes. He was an army man. Why do you ask?'

'It wouldn't be Captain Dane of the Munster Fusiliers, sir? Would it?'

'Yes, that was his Regiment. Why?'

'But I was his batman, sir. I was with him in France in 1916 when he got the bullet that killed him. Captain Dane, he died in my arms, sir.'

In Black

— *and* —

White

JAN MARK

Jenny Fielding is Mrs Sanderson, now. She has a husband, two daughters, Julia and Margery, and three grandchildren. On the sideboard in the living-room stand photographs of them all; daughters, sons-in-law, granddaughters, grandson. Every year Julia and Margery send new school photographs of Angus, Alice, and Rose. Mrs Sanderson arranges them on the sideboard and puts last year's photographs in her dressing-table drawer.

On the wall, above the sideboard, hangs Mr Sanderson's school photograph. It is black and white and a metre long, the whole school in it together. One after another Angus and Alice and Rose have asked, 'Grandpa, how was it done?' and Mr Sanderson explains that once upon a time all school photographs were like that, and had to be taken with a special camera. Everybody was arranged in a huge semicircle—there were seven hundred people at his school, and the camera, which was clockwork, slowly turned, panning from one end of the curve to the other. The real miracle is that, in the photograph, everyone is standing in a straight line while the building behind them looks curved. Grandpa tries to show them why, but they can never quite understand.

'You had to stand absolutely still,' Mr Sanderson says, 'because you could never be sure when the camera was pointing exactly at you.'

Angus and Alice and Rose love it when he gets to that point because they know what is coming next. Mr Sanderson is in the middle of the fourth row, looking very young and serious, with a surprising amount of hair, but at either end of the second row are the Schofeldt Triplets.

'Really, they were twins, Marcus and Ben,' Mr Sanderson tells them, 'and they were standing one each end of the row. When the camera got half-way round Ben left his place and ran along the back of the others, faster than the camera was moving, and went to stand beside Marcus at the other end. They got into a terrible row when they were found out, but we all thought they were heroes because we'd been forbidden to do it.'

Then Angus, Alice, and Rose look closely at Grandpa's school photograph to admire the three identical and heroic Schofeldt Twins, Ben at one end, Ben and Marcus at the other.

'Lots of school photographs had mysterious identical twins at each end,' Grandpa boasts, 'but I bet ours was the only one with triplets.'

'Did you have a school photograph, Granny?' Rose asks.

'I did once,' Mrs Sanderson says, vaguely, 'but I must have lost it.'

She hates lying, but if she told the truth about her school photograph no one would believe her anyway, so she pretends it is lost. But at the back of her dressing-table drawer, where Angus and Alice and Rose also lie, growing older and larger each year, is Jenny Fielding's school photograph, still rolled into a cylinder as it was on the day she first brought it home, forty years ago. She has never shown it to anyone since.

Jenny was thirteen, in the third year, when the notice was given out in assembly that the photographer was coming the following Monday. Miss Shaw, the form teacher, had a few words of her own to add when they returned to the classroom.

'You will all make sure that your uniforms are clean and pressed, that your hair is tidy—you'd better plait yours, Maureen Blake—and your shoes polished. I do not *care* if nobody can see your feet. There will be a rehearsal on Friday, so that each girl knows where she is to stand. And wherever you stand on Friday,' said Miss Shaw, fixing them with an iron gaze, 'you will stand on Monday. On these occasions there are always certain stupid people who imagine that it is amusing to run from one end of the line to the other in order to appear twice. Anyone who does that will be dealt with severely. Do I make myself clear?'

3a gazed back at her unblinkingly. Miss Shaw, as always, had made herself very clear. But in the back row Jenny's great friend

Margery Fletcher turned her head slightly and muttered to Jenny, 'I bet it will be one of us. I have a feeling.'

'Did you speak, Margery?' Miss Shaw enquired, knowing perfectly well.

'I just said I thought I might get my hair cut,' Margery said, pleasantly. 'For the photograph, you know.'

'An excellent idea,' Miss Shaw said. Margery's hair, like Jenny's, was wild and dark and curly. They were very alike in other ways, too; exactly the same height, short and stocky, and were often mistaken for each other by people who saw them misbehaving from a distance. Margery misbehaved far more frequently, and far more inventively, than Jenny, but when Jenny was falsely accused Margery always raised her hand and owned up. And on the rarer occasions when the mistake was in Jenny's favour, Jenny did the same. That was why they were best friends, faithful and true. They went everywhere together, near enough.

It had come as a surprise to no one when the announcement was made in assembly; bush telegraph had seen to that. Everybody had known for weeks that the photographer was due and some people even claimed—wrongly as it turned out— to know the date. So it was already public knowledge that after the rehearsal on Friday morning the lottery would take place. They had to wait until Friday to find out who would be in it.

Friday involved a great deal of standing about in a chilly damp wind on the lawn in front of the school. In the centre of the lawn stood a long curved row of eighty chairs, with a row of benches behind them and a row of tables behind that. One after another the classes stepped forward to take their places. On the chairs sat the sixth form with the teachers in the middle and the Headmistress in the very centre. In front of them the second years knelt upright, the most uncomfortable position of all, and right at the front sat the first years, cross-legged and trying not to show their knickers. Because they were only first years people thought that they were too young to care.

The third years stood on the grass behind the sixth form and staff, the fourth year stood behind them on the benches, and at the back the fifth forms teetered on the tables. Symmetry was all. The tallest in every group stood in the middle, the shortest at the sides, and so it was that Jenny and Margery found themselves facing each other across the grass at opposite ends of the third year, and Jenny was remembering what Margery had said last week: 'I bet it will be one of us.' There was a very good chance that it would be, one chance in four, but if it were, Jenny would be the one. Jenny was on the left, the end that the camera started from.

The entrants for the lottery met at the back of the sports pavilion after lunch; Jenny from the third year, one from the second year, one from the fourth and one from the fifth; all the left-hand tail-enders, except for the first year who were considered too young to be trusted, and the sixth who were above such things. Glenda Alcott, the fifth former, was there before them, holding her blue felt school hat in which lay four tightly folded pieces of paper.

'Now then,' Glenda said, 'three of these are blank and one carries the Black Spot. Whoever draws the Black Spot is the one who changes ends. As soon as the camera is pointing to the middle you leave your place and run round to the other end of the line. You know you'll get into a row afterwards. Are you prepared to risk it?'

The other three nodded solemnly.

'All right, then. Draw your papers.'

Madeline Enderby from the second year drew first, then Jenny, then Dawn Fuggle from the fourth, and that left one paper in the hat for Glenda and she took it out last of all.

Madeline, Dawn, and Glenda looked at each other before they looked at their papers, smiling but grim, as if they had been drawing lots to see who should go to the guillotine, but Jenny just stared at her folded paper, remembering what Margery had said: 'I bet it will be one of us. I have a feeling.' Margery had had feelings before, and they had come true. She had had a feeling before the carol concert last year, that she would be singing the descant in *Adeste Fideles*, and when Susan Beale lost her voice just before they were due to start it had been Margery who was called out to take her place. Then she had had a feeling about the geometry exam that everyone had been so worried about before Easter. 'I have a feeling there won't be an exam,' said Margery, who had done no revision, and on the morning that it was due to take place, Miss Ogden's briefcase, containing the papers, was stolen on the train.

'I have a feeling Cranmer House won't win the acting prize this year,' Margery said, the day before the drama competition, although Cranmer House were a dead cert, and sure enough, on the day, Cranmer went to pieces and fluffed their lines and missed their cues and the cup was awarded to Becket House. Margery and Jenny were in Becket.

Margery's feelings always seemed to involve misfortune for someone, Jenny sometimes reflected, but you couldn't blame Margery for that. *She* hadn't given Susan laryngitis, or nicked Miss Ogden's briefcase. Margery hadn't nobbled the entire cast of Cranmer's play.

'Open your papers,' Glenda said, and Jenny unfolded the little wad in her hand. She hardly needed to look; she knew that it would be her paper that bore the Black Spot.

'You can't back out now,' Dawn said, half envious, half relieved, when Jenny continued to stare at the paper in her palm.

'Remember what I said,' Glenda was admonishing her. 'Wait until the camera's half-way round in case you're still in shot, then run like hell.' Madeline gasped. She was only a second year. It seemed to her a very desperate thing that grown-up Glenda should say 'hell'.

'And another thing,' Glenda said. 'Don't tell anybody else who's won, except you, Jenny. You must appoint a liaison officer. If you're going to be feeble and come down with something at the last moment you must let us know before Monday lunchtime, so that we three can draw again.'

Jenny knew that there was no chance that she would come down with anything or Margery would have mentioned it, but she had to do what Glenda said, just in case. 'Will you be my liaison officer?' Jenny asked Margery, who showed no surprise when Jenny silently handed her the Black Spot.

'No need,' Margery said. 'If anything happens to you I'll run instead.'

'But you'd have to swap ends,' Jenny said. 'It doesn't work if you run the other way. You don't show up at all.'

'That won't be hard,' Margery said. 'People will think it's you anyway. They usually do. Actually,' she added, 'I have a feeling I may have to do it.'

'Why, am I going to drop dead before Monday?' Jenny snapped. Suddenly she felt that she had had enough of Margery and her feelings.

'Only joking,' Margery said, but Jenny had turned away with an angry flounce. During country dancing that afternoon, she chose Diana Sullivan for her partner, leaving Margery to the mercies of Galumphing Gertie the Games Mistress, who always stomped in enthusiastically to help out anyone who didn't have a partner, and at the end of the afternoon she went straight home alone instead of waiting for Margery who was in a different set for maths.

On Monday morning she made herself especially tidy, as demanded, for the photograph. Rumour had it that school photographs were always taken on Mondays so that even the scruffiest girls might look half-way presentable before they went downhill during the week.

Waiting in the form room for assembly they preened and checked each other out, even though there was the whole morning and lunch to get through before it was time for the photograph, so Jenny had only just noticed that Margery was not in the room before Miss Shaw appeared at the door and beckoned her out.

'Jenny, dear,' Miss Shaw said, as they stood in the corridor, 'I wanted a word with you before I told the others—I know Margery is a very special friend of yours.'

Jenny did not have feelings, not the way Margery did, but she knew what was coming.

'Margery had an accident yesterday,' Miss Shaw said. 'She was out for a drive with some family friends and the car door wasn't properly shut. Margery was thrown out into the road when they took a bend too sharply. She's in hospital. I'm afraid she's badly hurt.'

Jenny, excused assembly, went to sit in the cloakroom and listened to the swarming sound of rubber-shod feet as class after class converged upon the hall. The Headmistress must have made an announcement—perhaps they had all said a prayer for Margery's recovery—for at break the news was all round the school. Glenda Alcott came to find Jenny.

'You needn't run if you don't want to,' Glenda said, kindly. 'We'll understand.'

'I'll be all right,' Jenny said, 'Margery wouldn't want me to back out,' but she wasn't too bothered by what Margery would have wanted. All she knew was that if she had the photograph to worry about she might not have to think of Margery herself, lying in the hospital. 'A coma,' Miss Shaw had said. 'Severe head injuries.'

While they were all lining up after lunch, to go out on to the field, Glenda sought her out again.

'Listen,' she said, 'someone told me—someone who *knows*—' she added defiantly, 'that they do it twice, just in case anyone does run.'

'Margery had a feeling they'd do that,' Jenny said.

'The first time they don't run the film. Then if you leave your place you get caught and sent back and you don't dare try it again when they go for the take,' Glenda said. 'That's how they did it at my brother's school. They did it last time we had one here, too, but I didn't realize why. I was only a first year, then.'

If it had been Glenda alone who'd said it, Jenny would probably have doubted, and panicked, and spoiled her chance by running too soon, but as they stood there, tier upon tier, as they had on Friday, she looked across that great curve to the place where Margery ought to have been standing, and did not move. And Margery and Glenda had been right.

After the camera had swept round, and while they all stood there frozen and smirking, the little photographer blew his whistle, said, 'All right, ladies, let's do it once more, to make

sure,' and redirected the camera, on its tripod, towards Jenny's end of the line. He sounded his whistle again to warn them that he was ready to start and very slowly the camera began to turn a second time. Jenny thought how sinister it looked, clicking round on its plate, but the first time she had counted the seconds until it seemed to have reached the Headmistress, slap in the middle of the curve, and now, when the moment came, she took a step backwards, turned and began to run.

She hadn't thought before about what it would be like behind the curve. The backs of the fifth years, standing on their tables, reared eight feet above her, blotting out the sun; a palisade of legs, a swathe of skirts, a battlement of heads. The curve seemed endless, for she couldn't *see* the end of it, and the camera was so far ahead of her. In her mind's eye she could see that, the little black eye, inexorably turning, and she ran faster, racing her hidden adversary on the other side of the curve.

Three yards from the end of the line she slipped. The grass was damp where it had lain all day in the shade, her foot skidded from under her and, as she was off-balance already, leaning forward for the final effort, she fell flat, heavily, and lay there

winded, all the air slammed out of her lungs. She thought she was going to die, but suddenly she was able to breathe again and scrambled to her feet. But it was too late to run on. As she rose upright the wall of backs relaxed, there was a surge of muted laughter and conversation. The camera had got there first, the photographer had won and the photograph was over. Glenda Alcott, who had seen her leave and had, of course, been able to see also that she had not arrived at the far end, jumped down from the table and hurried round to find out what had happened.

'Did you fall? Bad luck. Hey, don't cry,' Glenda said, when she found Jenny weeping on the grass. Madeline and Dawn, the other tail-enders, were not so charitable.

'If you couldn't do it you might have said, and one of us could have run,' Madeline grumbled.

'I tried. I did try,' Jenny wept.

'Jenny has something on her mind,' Glenda said, severely, and the other two, remembering what it must be, became all at once very serious.

Jenny's mother came up to the school at the end of the afternoon, to meet Jenny and take her home. Jenny was far too old to be taken to and from school, but her mother had something to tell her. Margery had died at just after two o'clock, while they were having the photograph taken.

Everyone at school, girls and teachers, was kind and sympathetic to Jenny—until the photographs arrived, and then the storm broke, for there was Jenny, standing on the left-hand side of the picture, and there, in all her guilt, at the far end, was Jenny again, looking a little dishevelled and blurred, as though she had moved at the wrong moment.

'It isn't me,' Jenny kept saying.

'The truth, if you please,' said first Miss Shaw and then the Headmistress. 'Are you going to tell me that you didn't leave your place?'

'Yes, I did,' Jenny said. 'I did go, I did run round, but I never got there. I fell over.'

She was, as promised, severely dealt with; barred from this and banned from that, and everyone despised her for not admitting to what she had done, when the evidence was there in black and white, for anyone to see; except for three people. Glenda Alcott, Madeline Enderby, and Dawn Fuggle had all seen her leave her place, had all been watching the far end to see her arrive, and they alone knew that she had never got there. Lesley Wilson, the girl who was standing next but one at the end of the line and who had, on the day, been at the very end, to start with, said, 'Of course you were standing next to me. I felt your arm. Only I thought at first it was Margery—I mean, it should have been Margery, shouldn't it?'

Glenda borrowed a magnifying glass and they studied that indistinct little figure at the right-hand end of the photograph. 'It *could* be you,' she said, finally. What she didn't say, and what they were all thinking, was, 'It could be Margery.'

'She said if anything happened to me she'd be there in my place,' Jenny said. 'She had a feeling.'

This is why Mrs Sanderson keeps her school photograph in a drawer instead of hanging it on the wall beside her husband's. Even now, forty years later, she can't bring herself to explain.

The Warning

JOYCE MARSH

Philip stepped out briskly through the gathering dusk: if he was to be back at his hotel in time for dinner he would have to hurry. He shivered. The day had been overcast and chilly, and now, with the night coming on, it had turned quite cold—more like November than September, Philip thought gloomily.

For the first few days of his holiday the weather had not been kind, which was a pity, for this was a very special holiday. At eighteen, this was the first time that Philip had been away alone without either his parents or a school party. He had chosen to come to Devon because he had already spent many childhood holidays here, when the sun had always seemed to shine on blue seas and pale, white-gold sands.

But this was not the weather for lounging on the beach. Fortunately, however, this remote corner of Devon could offer a happy alternative in the lonely wilderness of its moors. Philip had already walked for miles, exploring the many little tracks and lanes which led off from the main road to meander through the wild heather before dropping down to the sea.

He zipped up his anorak against the night wind, which stung his cheeks as it whistled through the stiff gorse and mingled with the plaintive call of distant night birds. Before him, the road stretched blankly for miles, and Philip knew that he must walk at least as far as he could see before the welcoming lights of his hotel would come into sight.

A little way in the distance he could see the white gleam of a signpost pointing with one rigidly outstretched arm. Philip was immediately interested; here was yet another lane to explore—

tomorrow, perhaps. He came up to the signpost and stopped to read its directions.

'Pant . . .' The rest of the name had long since been worn away by wind and weather, and it was hardly surprising that no one had bothered to re-paint it, for the lane to 'Pant . . . something-or-other' was rough and so overgrown that it was almost non-existent. It seemed that few people these days travelled that way.

Philip was so intent upon his speculations as to the possible ending to a place name beginning 'Pant . . .' that the ferocious snarling of a dog somewhere behind him came as a sudden shock. Even as he spun round, Philip had a second to notice that the wind had dropped and, except for the snarling growls, an eerie hush had descended on the moor.

The dog was standing squarely in the middle of the road, barring his way. It was a spaniel—not a large dog, but, like all its breed, sturdily strong and muscular. It was staring at Philip with evil, white-ringed eyes, whilst its lips were lifted to show wicked, snarling teeth. Deep, rattling growls began in its throat and rose to a crescendo ending in a menacing bark.

Philip scowled. An evil-tempered brute like this should never be allowed out alone. He advanced on the dog, shouting and flapping his arms.

'Go-orn . . . get-out-of-it!'

The dog was not to be threatened off; it became even more menacing. Philip retreated slightly and looked around. There was no one in sight, and there was no house or village for miles. It was odd that such a beautiful little animal should be out on the moors alone, for it looked well cared for. The light rippled and gleamed on its long, silky ears and rich, golden-brown coat, whilst its domed, intelligent-looking head was highlighted by a flash of soft, pale blond hair.

Suddenly, with a little chill, Philip realized that he ought not to be able to see so much detail in this dusky light—but he could. It was almost as if the dog was . . . well, lit up from within. There was something uncanny about this dog, and Philip became anxious to slip past and be on his way as quickly as possible. He took a step first to one side and then the other, but the dog was not deceived by this manoeuvre. It moved with him. Then Philip saw the powerful muscles tense for a spring, and instinctively he braced himself to ward off the fearsome, snapping jaws. The dog was frighteningly quiet now, the light flickering along its rippling muscles as, with a powerful lunge, it launched itself at Philip. The dog sprang high. It was close enough for him to see the slimy strings of saliva drooling from its open jaws. He flung up an arm to protect his face . . . Then, suddenly, in mid-leap, it . . . vanished.

In all the stories Philip had read about people who encounter ghosts, they are described as searching for a natural explanation when their apparition disappears, but he knew at once that no normal body movement had carried the dog away. One second it was there, then . . . poof! . . . it was gone.

He was quite gratified to find that he was not in the least afraid. On the contrary, he was relieved. A real dog might have given him a nasty bite, but common sense told him that a ghost dog could do him no real harm.

Nevertheless, this was an eerie spot, and he had no mind to linger, so he set off again at a brisk jog-trot. As he ran, a smug little smile played about his lips. He was quite pleased to have joined the ranks of those who have actually seen a ghost. He imagined himself relating his experience and creating quite a stir in the office where he worked. However, if the story was to have its full impact it needed a bit of background. He promised himself that he would make a few local enquiries about that fierce little cocker spaniel.

The pleasant, purplish-grey dusk had darkened into gloomy night and, as he ran, Philip's light-hearted acceptance of his ghostly vision began to fade. He became increasingly nervous and oppressed by the eerie quietness of the moor. He could just make out the clumps of heather and gorse growing close to the roadside, but beyond that the moor seemed to drop away into black emptiness. He had the nerve-racking impression that he and the road along which he ran were silently suspended in a vast nothingness.

To add to his nervousness, he realized that he was not completely alone. Once or twice he glimpsed a vague shape moving through the bushes by the roadside, and keeping pace with him as he ran. Instinctively, Philip knew that it was the dog. He slowed his pace and took several deep breaths; he was surprised to find himself trembling, and he had to fight against an unreasoning urge to turn around and flee along the way he had come.

Anxiously he peered ahead into the distance and, to his relief, he could just make out the tiny little pin pricks of light which must be the village and the safe comfort of his hotel. Eager to be home, he went on again, but he had moved only a few paces when he was puzzled to see a faintly glowing shape lying in the roadway some little distance ahead.

His heart thumped with an inexplicable fear as he forced himself to walk towards the unknown thing which lay between him and safety. He came very close and was within a foot of it before he could recognize the box-like shape for what it was. Then he stopped dead, and a shuddering thrill of cold horror ran through him, for he found himself staring down at a gleaming, brand-new coffin.

Horribly abandoned, it lay there on the road with an unnatural radiance coming from the highly polished wood and finely wrought metal handles. A brass nameplate, ominously blank, was set into the lid, and carefully arranged along the top was a single, pure white lily. It was so close that, despite his horrified revulsion, Philip bent forward to touch it. The wood was curiously unresisting, and so cold that Philip felt as if his

fingers were sinking into slime. He tried to draw back, but his hand was suddenly seized from inside and tightly held by icy fingers.

For one horrifying minute the boy felt as if some awful dead thing was trying to use him to drag itself free of its grave. With a shuddering effort he freed his hand and leapt back. Then, with horrified fascination, he watched as the coffin began to disappear. It did not vanish instantly, as the dog had done, but slowly. With a kind of lingering reluctance it sank into the road. The last thing Philip saw was the flower, the fat, white lily, and then the surface closed over it.

Philip examined the roadway. There was no mark to show where the coffin had stood. A trembling fear began in the boy's knees and flooded through his body, then he began to run, and he did not stop until he could see the comforting welcome of the light flooding out through the open door of the hotel.

The next day Philip made a few tentative enquiries, but he could find no one who had heard of a dog haunting the main road over the moor. He discovered that the lane near the spot where he had seen the dog led to a place known locally as Pantacombe Bay, but his description of the ghostly spaniel was greeted with such amused disbelief that he could not bring himself to mention the second and more horrifying apparition.

He pretended to accept that the vanishing dog had been no more than a trick of the light, nevertheless he brooded on his experience. Fearfully, he told himself that the dog must have appeared as a kind of warning, and the coffin was the forecast of some frightful tragedy to come if he did not heed that warning. One thing was for sure, he decided, nothing was going to tempt him to walk along that particular road again.

However, as his holiday neared its end, he found that he had explored as much of the district as he could without travelling on the main road. The weather was still too chilly for sporting on the beach, and he was beginning to feel bored. So, when he heard of an old fishing village whose inhabitants had long ago been driven away by the erosion of the sea, he could not resist going to explore, even though it meant walking over the haunted road. In the crowded comfort of the hotel lounge he made his plans and confidently told himself that his chances of being haunted for the second time were very slight.

Nevertheless, he set off early, promising himself that he would be sure to start for home in good time to be safely back before dusk. But, as the poet tells us, even the best laid plans 'gang aft a-gley'.

The village was a long way off, and it was difficult to find; he did not arrive until well into the afternoon, and by the time he had explored the ruined, tide-washed cottages, the bright glitter of the sun had darkened to a glowing orange. Even as he started out for home, he knew that he could not possibly pass the lane to Pantacombe Bay before dusk. Nevertheless, he hurried, in a desperate bid to beat the sun as it fell down towards the edge of the moor. It was a hopeless race and one that he could not win; his eyes were already straining through the purple-grey dusk as he saw the signpost pointing to Pantacombe Bay.

With sinking dread he felt once again that hushed stillness descend about him. Then he saw the dog. It was in the middle of the road, not snarling or growling yet—just waiting! By its weird, inner light he could see the rich golden coat and the deep brown eyes gazing fixedly towards him. The dog lifted its snout to the sky and sent out a low, whining wail. Philip felt the hairs prickle on his scalp, and his hands became moist with the sweat of his fear. Instinctively, he knew that if the dog was there it was a warning that the other thing would be there, too—lying in the road waiting to deliver its horrible prophecy of some mysterious doom.

He walked on—there was nothing else to do; somehow he had to find the courage to pass the dog and then the dreadful, abandoned coffin. As he drew nearer, the spaniel lifted its lips in a snarl and the growling began deep in its throat.

Philip's little store of courage deserted him. He sprang back, looking wildly around. Surely other people used this road; there had been cars enough this morning—why could not one come along now? But the moor was deserted and eerily quiet. Then his eye fell upon the lane to Pantacombe—of course, why had he not thought of it before? The lane was rough, and heaven knew where it led, but it was a means of escape. He darted down it, running as fast as he could over the rutted surface. Once he looked behind, and the dog was following, gliding smoothly and silently, yet warily, as if challenging him to turn back.

About fifty yards down the lane he glimpsed the thatched roof of a cottage nestling amidst a solitary clump of trees, and to his enormous relief there were cheerful lights blazing out of the windows. He ran towards it, but as his hand reached for the gate he saw the dog bound forward. Its stumpy little tail was vibrating merrily, and it bounced slightly, as dogs do when they are pleased. The spaniel was no more than a foot or two away when, just as it had done before, the little dog abruptly vanished.

At once, all the little night sounds returned; the evening breeze carried the faint tang of the sea as it rustled pleasantly through the trees. Everything now seemed so normal that Philip was suddenly ashamed of his previous panic. He took his hand away from the gate and decided to go on, not back to the road—his newly returned courage was not quite equal to that—but on down the lane to Pantacombe.

The rough track narrowed and became even more overgrown as it dropped steeply down towards the sea. If the village of Pantacombe had ever existed, there was no stick or stone left of it now. The road eventually led on to the cliff edge, where it curved around, following the line of the bay, before it turned back across the moor to rejoin the main road a hundred yards or so beyond Philip's hotel.

It was more than three years before Philip returned to Devon. He was that much older, and the pressures of business did not allow him so much time for walking. Therefore, the next time he took that familiar road across the moor, he was driving—and in some haste, for he had a dinner appointment. With a twinge of apprehension he realized that it was dusk, and fading into night when he saw, once again, the tall signpost still rigidly pointing to the non-existent Pantacombe. Nervously he tried to push aside the memory of the snarling dog which had twice barred his way at this spot.

He desperately wanted to press down his foot and speed past, but he could not. Some compulsion stronger than his own will forced him to brake hard and bring his car to a slightly screeching halt. He half expected to see the snarling dog still standing there, but the road was empty. It stretched before him, blankly inviting. Suddenly, and with an unreasoning surge of relief, he knew what he must do! Wrenching at the wheel, he sent his car leaping off down the lane, which was now even more overgrown and deeply rutted.

The thatched roof of the cottage soon came into sight, and Philip, curious to know if it was still occupied, slowed down almost to a stop. And there, sitting by the gate, was the dog—the same beautiful little golden cocker spaniel. It was gazing expectantly up the lane, as if it was waiting patiently just for Philip to appear. As soon as the car slowed, the dog trotted towards it and jumped up to look in through the side window. With some relief, Philip noticed that the scrape of its feet against the car door sounded reassuringly real.

For a long moment the dog gazed at Philip with eyes full of dumb appeal, then it jumped down and trotted back to the cottage. After a few steps, however, the dog stopped, looked back at the car and gave a short, sharp bark. There was no mistaking its meaning—the animal was asking Philip to follow.

Mystified, the young man allowed himself to be led up to the cottage door. It was not locked, and opened at his touch. Inside, the house was heavy with the scent of flowers and the fresh, clean smell of wax polish, but there was a hushed stillness which told Philip that the cottage was empty. Still mystified, he paused. It was embarrassing to find himself here as an uninvited stranger in a stranger's home. But the dog, it seemed, had no time for the niceties of human behaviour. He gave sharp, impatient barks and, in his own unmistakable way, demanded that Philip follow him to the back of the house and into the kitchen.

The light was burning in here, but at first glance this room, too, appeared to be empty. The dog now became frantic. It seized Philip's trouser leg and began to tug at it with all his strength. Philip was pulled further into the room, and it was then he saw the girl lying on the floor.

She was about his own age and very pretty except for the deathly pallor of her face. She was lying still—so still that it was impossible to see if she was breathing. One arm was flung out, and a little fountain of bright red blood was bubbling out from a long gash in her wrist.

Philip had a moment to notice the butcher's knife on the floor beside her and the piece of frozen meat on the table. The girl must have been trying to cut the meat when the knife slipped. He dropped to his knees beside her and his fingers probed her upper arm for the heavy, throbbing pulse which was pumping away her life's blood. With his other hand he ripped off his tie, and when he found the pulse he tied the tie tightly around her arm at that point. There was a spoon on the table, and he thrust it through the tie and used it as a handle to twist and tighten his improvised tourniquet.

To his relief, Philip saw the heavy drain of blood ease to a thin trickle, but there was still no time to lose. The girl was scarcely breathing, and in no more than fifteen minutes the tourniquet must be removed. Bending down he scooped her up in his arms; she was very slight and her weight scarcely hampered him at all as he raced back to his car and dumped her a trifle unceremoniously on to the back seat. The dog was plainly not going to be separated from his mistress; he scrambled into the car beside Philip and stood on the seat, gazing down at her with pathetic helplessness.

Philip's car was fast, and he knew how to push it to its limits. Thankfully, his memory served him well, for he remembered every twist and turn in the road as he raced across the moor and down the steep road to the nearest town, where he knew there was a hospital.

It was a hair-raising ride, but in little more than ten minutes Philip was able to hand over his charge into the calm, efficient care of doctors and nurses. Now there was nothing more he could do except wait, and he waited a long time before someone came to tell him that the girl would live. She was very weak from loss of blood, but she was young and strong and would soon recover.

Philip had been in the hospital for so long that he had forgotten the dog, and it came as a surprise when he returned to the car and found the spaniel still patiently waiting. With a slight hesitancy, Philip put out his hand and for the first time touched the soft hair on the dog's head. He ran his hand over the spaniel's flanks and was almost surprised to feel the warm throb of life.

'Well, old boy, your mistress is going to be all right, but it looks as if I'm going to be stuck with you—at least until tomorrow.'

The swift wagging of the short, stumpy tail seemed to indicate that the dog had no objection to this arrangement.

It was very early next morning when Philip returned to the cottage. Ostensibly he went to take the dog home, but, as he readily admitted to himself, his real interest was in the very pretty girl.

The cottage door was opened by a plump, middle-aged lady with a flat, country face. The dog hurled himself upon her, excitedly nibbling her apron and leaping up to lick her face.

'Well, bless me, if it isn't our Mufti! Then you must be the young gentleman as saved our girl. Well, do come in, sir. We just don't know how to thank you for what you did, nor give praise enough for the miracle which sent you down our lane last night. Why, yours must be the first car that's come down here for weeks. Then, to beat all, you stopped and came in—what blessed chance made you do that?'

'It was the dog—Mufti. He made me come into your cottage.'

'Oh, you dear thing!' She bent down to give Mufti a bone-crushing hug. 'So you helped to save your mistress . . . and to think we nearly took you with us. You see, sir—' She turned back to Philip. 'Father and me were away to market yesterday, and we usually take Mufti with us to mind the van, only yesterday he wouldn't come. The little rascal ran off and hid somewhere, so we went without him. What a shock it was when we came back and found the police waiting to tell us what had happened. Lawd-a-mercy, my blood runs cold to think on it. There's no doubt our Lily would have been in her coffin be now if you hadn't come along.'

Lily. The single white flower lying along the lid of that ghostly coffin—that had been a lily. Dimly, Philip began to understand. The coffin might have been for the girl, Lily, but it had not been a prophecy of inevitable doom, it had been a warning of a tragedy which could be averted.

'It's really uncanny,' Lily's mother was saying, 'the way that dog just would not come wi' us. It's as if he knew he'd be needed here.'

It is even more uncanny than you know, Philip thought as he gazed down at the excited Mufti.

'Have you had Mufti long?' he asked on an impulse.

'No, not long. He's Lily's dog, really. We used to have another golden cocker, only he died, very sudden like, about three years ago. Lily was broken-hearted, and she gave us no peace until we got her another just like him.'

So it was not this real, living dog, but the spirit of a dead dog which had lingered on, waiting for a stranger who would heed his warning and remember it.

When Lily was well enough to come home, Philip went to the cottage very often, and they became friends, but somehow he could never bring himself to tell her of the vision which, three years before, had warned him that one day he was to be the means of saving her life.

Years later, when Lily and Philip were comfortably married, they would sit before their fire with an ageing Mufti snoring at their feet. Sometimes they would talk of the accident which had brought them together, then Mufti would open one lazy eye to gaze up at Philip. There was a deep, indefinable expression in that one eye as the old dog slowly lifted his lip in a silent snarl—which was very odd, for Mufti never snarled at anyone—least of all his beloved master.

John Charrington's Wedding

E. NESBIT

No one ever thought that May Forster would marry John Charrington; but he thought differently, and things which John Charrington intended had a queer way of coming to pass. He asked her to marry him before he went up to Oxford. She laughed and refused him. He asked her again next time he came home. Again she laughed, tossed her dainty blonde head, and again refused. A third time he asked her; she said it was becoming a confirmed bad habit, and laughed at him more than ever.

John was not the only man who wanted to marry her: she was the belle of our village *coterie*, and we were all in love with her more or less; it was a sort of fashion, like heliotrope ties or Inverness capes. Therefore we were as much annoyed as surprised when John Charrington walked into our little local Club—we held it in a loft over the saddler's, I remember—and invited us all to his wedding.

'Your wedding?'

'You don't mean it?'

'Who's the happy fair? When's it to be?'

John Charrington filled his pipe and lighted it before he replied. Then he said:

'I'm sorry to deprive you fellows of your only joke—but Miss Forster and I are to be married in September.'

'You don't mean it?'

'He's got the mitten again, and it's turned his head.'

'No,' I said, rising, 'I see it's true. Lend me a pistol someone—or a first-class fare to the other end of Nowhere. Charrington has bewitched the only pretty girl in our twenty-mile radius.

Was it mesmerism, or a love-potion, Jack?'

'Neither, sir, but a gift you'll never have—perseverance—and the best luck a man ever had in this world.'

There was something in his voice that silenced me, and all chaff of the other fellows failed to draw him further.

The queer thing about it was that when we congratulated Miss Forster, she blushed and smiled and dimpled, for all the world as though she were in love with him, and had been in love with him all the time. Upon my word, I think she had. Women are strange creatures.

We were all asked to the wedding. In Brixham everyone who was anybody knew everybody else who was any one. My sisters were, I truly believe, more interested in the *trousseau* than the bride herself, and I was to be best man. The coming marriage was much canvassed at afternoon tea-tables, and at our little Club over the saddler's, and the question was always asked: 'Does she care for him?'

I used to ask that question myself in the early days of their engagement, but after a certain evening in August I never asked it again. I was coming home from the Club through the churchyard. Our church is on a thyme-grown hill, and the turf about it is so thick and soft that one's footsteps are noiseless.

I made no sound as I vaulted the low lichened wall, and threaded my way between the tombstones. It was at the same instant that I heard John Charrington's voice, and saw Her. May was sitting on a low flat gravestone, her face turned towards the full splendour of the western sun. Its expression ended, at once and for ever, any question of love for him; it was transfigured to a beauty I should not have believed possible, even to that beautiful little face.

John lay at her feet, and it was his voice that broke the stillness of the golden August evening.

'My dear, my dear, I believe I should come back from the dead if you wanted me!'

I coughed at once to indicate my presence, and passed on into the shadow fully enlightened.

The wedding was to be early in September. Two days before I had to run up to town on business. The train was late, of course, for we are on the South-Eastern, and as I stood grumbling with my watch in my hand, whom should I see but John Charrington and May Forster. They were walking up and down the unfrequented end of the platform, arm in arm, looking into each other's eyes, careless of the sympathetic interest of the porters.

Of course I knew better than to hesitate a moment before burying myself in the booking-office, and it was not till the train drew up at the platform, that I obtrusively passed the pair and took the corner in a first-class smoking-carriage. I did this with as good an air of not seeing them as I could assume. I pride myself on my discretion, but if John were travelling alone I wanted his company. I had it.

'Hallo, old man,' came his cheery voice as he swung his bag into my carriage. 'Here's luck; I was expecting a dull journey!'

'Where are you off to?' I asked, discretion still bidding me turn my eyes away, though I saw, without looking, that hers were red-rimmed.

'To old Branbridge's,' he answered, shutting the door and leaning out for a last word with his sweetheart.

'Oh, I wish you wouldn't go, John,' she was saying in a low, earnest voice. 'I feel certain something will happen.'

'Do you think I should let anything happen to keep me, and the day after tomorrow our wedding day?'

'Don't go,' she answered, with a pleading intensity which would have sent me on to the platform. But she wasn't speaking to me. John Charrington was made differently; he rarely changed his opinions, never his resolutions.

He only stroked the little ungloved hands that lay on the carriage door.

'I must, May. The old boy's been awfully good to me, and now he's dying I must go and see him, but I shall come home in time for—' the rest of the parting was lost in a whisper and in the rattling lurch of the starting train.

'You're sure to come?' she spoke as the train moved.

'Nothing shall keep me,' he answered; and we steamed out. After he had seen the last of the little figure on the platform he leaned back in his corner and kept silence for a minute.

When he spoke it was to explain to me that his godfather, whose heir he was, lay dying at Peasmarsh Place, some fifty miles away, and had sent for John, and John had felt bound to go.

'I shall be surely back tomorrow,' he said, 'or, if not, the day after, in heaps of time. Thank Heaven, one hasn't to get up in the middle of the night to get married nowadays!'

'And suppose Mr Branbridge dies?'

'Alive or dead I mean to be married on Thursday!' John answered, lighting a cigar and unfolding *The Times*.

At Peasmarsh station we said 'Goodbye', and he got out, and I saw him ride off; I went on to London, where I stayed the night.

When I got home the next afternoon, a very wet one, by the way, my sister greeted me with:

'Where's Mr Charrington?'

'Goodness knows,' I answered testily. Every man, since Cain, has resented that kind of question.

'I thought you might have heard from him,' she went on, 'as you're to give him away tomorrow.'

'Isn't he back?' I asked, for I had confidently expected to find him at home.

'No, Geoffrey,' —my sister Fanny always had a way of jumping to conclusions, especially such conclusions as were least favourable to her fellow-creatures— 'he has not returned, and, what is more, you may depend upon it he won't. You mark my words, there'll be no wedding tomorrow.'

My sister Fanny has a power of annoying me which no other human being possesses.

'You mark my words,' I retorted with asperity, 'you had better give up making such a thundering idiot of yourself. There'll be more wedding tomorrow than ever you'll take the first part in.' A prophecy which, by the way, came true.

But though I could snarl confidently to my sister, I did not feel so comfortable when late that night, I, standing on the doorstep of John's house, heard that he had not returned. I went home gloomily through the rain. Next morning brought a brilliant blue sky, gold sun, and all such softness of air and beauty of cloud as go to make up a perfect day. I woke with a vague feeling of having gone to bed anxious, and of being rather averse to facing that anxiety in the light of full wakefulness.

But with my shaving-water came a note from John which relieved my mind and sent me up to the Forster's with a light heart.

May was in the garden. I saw her blue gown through the hollyhocks as the lodge gates swung to behind me. So I did not go up to the house, but turned aside down the turfed path.

'He's written to you too,' she said, without preliminary greeting, when I reached her side.

'Yes, I'm to meet him at the station at three, and come straight on to the church.'

Her face looked pale, but there was a brightness in her eyes, and a tender quiver about the mouth that spoke of renewed happiness.

'Mr Branbridge begged him so to stay another night that he had not the heart to refuse,' she went on. 'He is so kind, but I wish he hadn't stayed.'

I was at the station at half-past two. I felt rather annoyed with John. It seemed a sort of slight to the beautiful girl who loved him, that he should come as it were out of breath, and with the dust of travel upon him, to take her hand, which some of us would have given the best years of our lives to take.

But when the three o'clock train glided in, and glided out again having brought no passengers to our little station, I was more than annoyed. There was no other train for thirty-five minutes; I calculated that, with much hurry, we might just get to the church in time for the ceremony; but, oh, what a fool to miss that first train! What other man could have done it?

That thirty-five minutes seemed a year, as I wandered round the station reading the advertisements and the time-tables, and the company's bye-laws, and getting more and more angry with John Charrington. This confidence in his own power of getting everything he wanted the minute he wanted it was leading him too far. I hate waiting. Everyone does, but I believe I hate it more than anyone else. The three thirty-five was late, of course.

I ground my pipe between my teeth and stamped with impatience as I watched the signals. Click. The signal went down. Five minutes later I flung myself into the carriage that I had brought for John.

'Drive to the church!' I said, as someone shut the door. 'Mr Charrington hasn't come by this train.'

Anxiety now replaced anger. What had become of the man? Could he have been taken suddenly ill? I had never known him have a day's illness in his life. And even so he might have telegraphed. Some awful accident must have happened to him. The thought that he had played her false never—no, not for a moment—entered my head. Yes, something terrible had happened to him, and on me lay the task of telling his bride. I almost wished the carriage would upset and break my head so that someone else might tell her, not I, who—but that's nothing to do with this story.

It was five minutes to four as we drew up at the churchyard gate. A double row of eager onlookers lined the path from lychgate to porch. I sprang from the carriage and passed up between them. Our gardener had a good front place near the door. I stopped.

'Are they waiting still, Byles?' I asked, simply to gain time, for of course I knew they were by the waiting crowd's attentive attitude.

'Waiting, sir? No, no, sir; why, it must be over by now.'

'Over! Then Mr Charrington's come?'

'To the minute, sir; must have missed you somehow, and, I say, sir,' lowering his voice, 'I never see Mr John the least bit so afore, but my opinion is he's been drinking pretty free. His clothes was all dusty and his face like a sheet. I tell you I didn't like the looks of him at all, and the folks inside are saying all sorts of things. You'll see, something's gone very wrong with Mr John, and he's tried liquor. He looked like a ghost, and in he went with his eyes straight before him, and with never a look or a word for none of us: him that was always such a gentleman!'

I had never heard Byles make so long a speech. The crowd in the churchyard were talking in whispers and getting ready rice and slippers to throw at the bride and bridegroom. The ringers were ready with their hands on the ropes to ring out the merry peal as the bride and bridegroom should come out.

A murmur from the church announced them; out they came. Byles was right. John Charrington did not look himself. There was dust on his coat, his hair was disarranged. He seemed to have been in some row, for there was a black mark above his eyebrow. He was deathly pale. But his pallor was not greater than that of the bride, who might have been carved in ivory— dress, veil, orange blossoms, face and all.

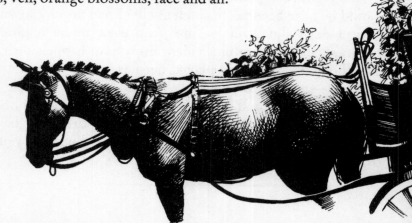

As they passed out the ringers stooped—there were six of them—and then, on the ears expecting the gay wedding peal, came the slow tolling of the passing bell.

A thrill of horror at so foolish a jest from the ringers passed through us all. But the ringers themselves dropped the ropes and fled like rabbits out into the sunlight. The bride shuddered, and grey shadows came about her mouth, but the bridegroom led her on down the path where the people stood with the handfuls of rice; but the handfuls were never thrown, and the wedding-bells never rang. In vain the ringers were urged to remedy their mistake: they protested with many whispered expletives that they would see themselves further first.

In a hush like the hush in the chamber of death the bridal pair passed into their carriage and its door slammed behind them.

Then the tongues were loosed. A babel of anger, wonder, conjecture from the guests and spectators.

'If I'd seen his condition, sir,' said old Forster to me as we drove off, 'I would have stretched him on the floor of the church, sir, by Heaven I would, before I'd have let him marry my daughter!'

Then he put his head out of the window.

'Drive like hell,' he cried to the coachman; 'don't spare the horses.'

He was obeyed. We passed the bride's carriage. I forbore to look at it, and old Forster turned his head away and swore. We reached home before it.

We stood in the hall doorway, in the blazing afternoon sun, and in about half a minute we heard the wheels crunching the gravel. When the carriage stopped in front of the steps old Forster and I ran down.

'Great Heaven, the carriage is empty! And yet—'

I had the door open in a minute, and this is what I saw—

No sign of John Charrington; and of May, his wife, only a huddled heap of white satin lying half on the floor of the carriage and half on the seat.

'I drove straight here, sir,' said the coachman, as the bride's father lifted her out; 'and I'll swear no one got out of the carriage.'

We carried her into the house in her bridal dress and drew back her veil. I saw her face. Shall I ever forget it? White, white and drawn with agony and horror, bearing such a look of terror as I have never seen since except in dreams. And her hair, her radiant blonde hair, I tell you it was white like snow.

As we stood, her father and I, half mad with the horror and mystery of it, a boy came up the avenue—a telegraph boy. They brought the orange envelope to me. I tore it open.

Mr Charrington was thrown from the dogcart on his way to the station at half-past one. Killed on the spot!

And he was married to May Forster in our parish church at *half-past three*, in the presence of half the parish.

'I shall be married, dead or alive!'

What had passed in that carriage on the homeward drive? No one knows—no one will ever know. Oh, May! oh my dear!

Before a week was over they laid her beside her husband in our little churchyard on the thyme-covered hill—the churchyard where they had kept their love-trysts.

Thus was accomplished John Charrington's wedding.

Dead Call

WILLIAM F. NOLAN

Len had been dead for a month when the phone rang. Midnight. Cold in the house and me dragged up from sleep to answer the call. Helen gone for the weekend. Me, alone in the house. And the phone ringing . . .

'Hello.'

'Hello, Frank.'

'Who is this?'

'You know *me*. It's Len . . . Len Stiles.'

Cold. Deep and intense. The receiver dead-cold metal in my hand.

'Leonard Stiles died four weeks ago.'

'Four weeks, three days, two hours and twenty-seven minutes ago—to be exact.'

'I want to know who you are!'

A chuckle. The same dry chuckle I'd heard so many times. 'C'mon, ole buddy—after twenty years. Hell, you *know* me.'

'This is a damned poor joke!'

'No joke, Frank. You're there, alive. And I'm here, dead. And you know something, ole buddy? I'm really *glad* I did it.'

'Did . . . what?'

'Killed myself. Because . . . death is just what I hoped it would be: beautiful . . . grey . . . quiet. No pressures.'

'Len Stiles' death was an accident . . . a concrete freeway barrier . . . His car—'

'I *aimed* my car for that barrier. Pedal to the floor. Doing almost a hundred when I hit . . . No accident, Frank.' The voice cold . . . Cold. 'I *wanted* to be dead . . . and no regrets.'

I tried to laugh, make light of this—matching his chuckle with my own. 'Dead men don't use telephones.'

'I'm not really using a phone, not in a physical sense. It's just that I chose to contact you this way. You might say it's a matter of psychic electricity. As a detached spirit, I'm able to align my cosmic vibrations to match the vibrations of this power line. Simple, really.'

'Sure. A snap. Nothing to it.'

'Naturally you're sceptical. I expected you to be. But . . . listen carefully to me, Frank.'

And I listened—with the phone gripped in my hand in that cold night house—as the voice told me things that *only* Len could know . . . intimate details of shared experiences extending back through two decades. And when he'd finished I was certain of one thing:

He *was* Len Stiles.

'But how . . . I still don't . . .'

'Think of this phone as a medium—a line of force through which I can bridge the gap between us.' The dry chuckle again. 'Hell, you gotta admit it beats holding hands around a table in the dark—yet the principle is the same.'

I'd been standing by my desk, transfixed by the voice. Now I moved behind the desk, sat down, trying to absorb this dark miracle. My muscles were wire-taut, my fingers cramped about the metal receiver. I dragged in a slow breath, the night dampness of the room pressing at me. 'All right . . . I don't believe in ghosts, don't pretend to understand any of this, but . . . I'll accept it. I *must* accept it.'

'I'm glad, Frank—because it's important that we talk.' A long moment of hesitation. Then the voice, lower now, softer. 'I *know* how lousy things have been, ole buddy.'

'What do you mean?'

'I just know how things are going for you. And . . . I want to help. As your friend. I want you to know that I understand.'

'Well . . . I'm really not—'

'You've been feeling bad, haven't you? Kind of *down* . . . right?'

'Yeah. A little, I guess.'

'And I don't blame you. You've got reasons. Lots of reasons . . . For one, there's your money problem.'

'I'm expecting a raise. Cooney promised me one—within the next few weeks.'

'You won't get it, Frank. I *know*. He's lying to you. Right now, at this moment, he's looking for a man to replace you at the company. Cooney's planning to fire you.'

'He never liked me. We never got along from the day I walked into that office.'

'And your wife . . . All the arguments you've been having with her lately . . . It's a pattern, Frank. Your marriage is all over. Helen's going to ask you for a divorce. She's in love with another man.'

'Who, dammit? What's his name?'

'You don't know him. Wouldn't change things if you did. There's nothing you can do about it now. Helen just . . . doesn't love you any more. These things happen to people.'

'We've been drifting apart for the last year . . . But I didn't know why. I had no idea that she—'

'And then there's Jan. She's back on it, Frank. Only it's worse now. A lot worse.'

I knew what he meant—and the coldness raked along my body. Jan was nineteen, my oldest daughter, and she'd been into drugs for the past three years. But she'd promised to quit.

'What do you know about Jan? Tell me!'

'She's into the heavy stuff, Frank. She's hooked bad. It's too late for her.'

'What the hell are you saying?'

'I'm saying she's lost to you . . . She's rejected you, and there's no reaching her. She *hates* you . . . Blames you for everything.'

'I won't *accept* that kind of blame! I did my best for her.'

'It wasn't enough, Frank. We both know that. You'll never see Jan again.'

The blackness was welling within me, a choking wave through my body.

'Listen to me, ole buddy . . . things are going to get worse, not better. I know. I went through my own kind of hell when I was alive.'

'I'll . . . start over. Leave the city. Go East, work with my brother in New York.'

'Your brother doesn't want you in his life. You'd be an intruder . . . an alien. He never writes you, does he?'

'No, but that doesn't mean—'

'Not even a card last Christmas. No letters or calls. He doesn't *want* you with him, Frank, believe me.'

And then he began to tell me other things. He began to talk about middle age, and how it was too late now to make any kind of new beginning. He spoke of disease . . . loneliness . . . of rejection and despair. And the blackness was complete.

'There's only one real solution to things, Frank—just *one*. That gun you keep in your desk upstairs. Use it, Frank. Use the gun.'

'I couldn't do that.'

'But why not? What other choice have you got? The solution is *there*. Go upstairs and use the gun. I'll be waiting for you afterwards. You won't be alone. It'll be just like the old days . . . We'll be together . . . death is beautiful . . . Use the gun, Frank . . . The gun . . . Use the gun . . . The gun . . . The gun . . .'

I've been dead for a month now, and Len was right. It's fine here. No pressures. No worries. Grey and quiet and beautiful . . .

I know how lousy things have been going for you. And they won't get any better.

Isn't that your phone ringing?

Better answer it.

It's important that we talk.

Caves in Cliffs

JOSH PACHTER

T hat's it for tonight,' Jack Farmer told his class of enlisted men and lower-echelon officers and dependant wives and husbands. 'See you next week. Don't forget your journals are due.'

He put away his notes and attendance sheet, slung his leather totebag over his shoulder and left the classroom.

Another day, he sighed contentedly, as the door swung shut behind him; *another drachma*.

He was a term-appointed instructor in the University of Maryland's European Division, teaching three or four eight-week semesters a year on American military bases in England, West Germany, Holland, Spain and—now—Greece. It was an idyllic job, with small groups of dedicated, motivated students meeting four evenings a week for a couple of hours per session; the pay was adequate if not spectacular, the schedule was hard to beat, and the university paid his travel expenses from place to place. He was in his third year with the programme, and he couldn't think of much else he would rather be doing with his life.

Humming merrily to himself, he walked out to his battered but trusty old Beetle and tossed his bag in the back seat and coaxed the engine to life.

He drove past the AFRTS station and the commissary and, with a wave to the gate guard, off the base. Two hundred yards down the road he turned off onto the narrow dirt trail that led to the beach, flicked on his brights and jounced forward slowly, playing his nightly game of connect-the-potholes.

Home, he checked the laundry for toothmarks—none; he'd hung it, for once, out of reach of the goats—pulled it from the line and let himself into his apartment.

Too warm for a fire, he decided. That was lucky, as he hadn't yet taken the time to go through the stack of driftwood he'd collected that afternoon to separate out the tarry pieces.

He kicked off his sandals, tossed his sweat-stained T-shirt at the bathroom hamper, poured himself a glass of the nameless local wine the neighbours kept plying him with, and spread out his roadmap on the living-room table.

It was Thursday night, and he had plans to make for the weekend that stretched languorously out before him.

His first weekend on Crete he'd stayed close to home, getting to know Iráklion on Friday and Saturday and paying what he'd expected to be a courtesy call to the what-he'd-expected-to-be-vastly-overrated Minoan ruins at Knossos on Sunday. Sir Arthur Evans' reconstruction had proven to be absolutely fascinating, though, and he'd wound up spending almost five hours there, then hurrying back to Iráklion to examine the Minoan relics in the archaeological museum while the images of the ancient city were still fresh in his mind. Then, on Monday morning, he'd gone up to the Plains of Lassíthi—and he would never forget that one awesome moment of cresting the final mountain peak to see, spread out beneath him, the incredible, other-worldly sight of ten thousand snow-white windmills spinning gently in the breeze.

The second weekend he'd headed west, to the charming fishing centres of Réthymnon and Khaniá, to drink beer and eat fresh squid and chatter away the days with French and Dutch tourists around the similar, yet individual, harbours.

The third weekend he'd gone east, skirting the ritzy resort town of Agios Nikólaos and arrowing straight out to Vai, where he'd lain on cool sands under the shade of Europe's only palm forest.

Iráklion, Knossos, and Lassíthi in the north, Réthymnon and Khaniá to the west, Vai to the east.

This weekend, then, it was time to take off for the south.

He pored over his map, examining the long stretch of Crete's southern coast carefully and waiting for inspiration to strike.

Halfway across the island, next to a village called Matala, a red star and a number appeared on the map, indicating the location of a sight worth seeing. He checked the reference guide on the reverse of the sheet, and by the corresponding number he found one short phrase, three simple words: 'Caves in cliffs.'

Caves in cliffs, he thought happily. *Of course!*

'*Matala*', Jack Farmer decided, before he had been there half an hour, *must be Greek for 'Shangri-la'.*

He had arrived early that afternoon, following a leisurely drive south across the waist of Crete, with stops only for lunch and a quick look around the ruins of Phaestos.

The village of Matala was practically non-existent: a square, a single street, another square. But the first square was ringed with charming outdoor cafés, and the street was covered over with a cloth awning to keep out the burning sun and lined with colourful vegetable stalls and tasteful souvenir stands and displays of local crafts, and around the second, smaller square

were three delightful *tavernas*, each decorated with hanging fishnets and outrageous murals of mythological creatures of the sea—and the whole was set on a rise overlooking a crescent of immaculate, deserted beach, and the shimmering mystery of the Mediterranean stretching all the way to Africa, and, of course, the cliffs.

One cliff, really, rising from the far end of the beach, a powerful structure of bright tan stone studded amazingly with row upon row of tiny black mouths, hundreds of them, reaching from twenty feet above the surface of the water up to the very top of the cliff: the caves.

No one he asked seemed to know how long they had been there, whether they were entirely man-made or if, originally, nature had had a hand in their construction. But he found out that, when the Aquarian Age had dawned, back in the '60s, an avalanche of backpacking, guitar-playing, pot-smoking hippies had discovered Matala's 'caves in cliffs', had enlarged them and extended them and covered their walls with mindblowing psychedelic design, had turned the hive of individual stone compartments into a commune which drew converts from all over Europe and as far away as the United States. Some of the flower children had stayed only a few weeks before rolling up their sleeping-bags and moving on; many had been there for months, though, and even years, until the complaints about nudity and drugs reached the ears of the Greek government at last, and the cave-dwellers were evicted and a chain-link fence put up between the beach and the cliff.

And Matala had gone back to sleep, and was sleeping still, a slumbering lassitude set apart from the rest of the world, untroubled by progress or current events or change. The toothless old women dressed in shapeless black, mourning long-dead fathers and brothers, husbands and sons. The old men let their hair and beards grow wild, and sat around the quiet cafés alone or in pairs, sipping silently at their tumblers of fiery *ouzo*. They sold vegetables to each other and trivia to the occasional tourists. They tended their gardens and goats and chickens and children, and their little girls grew older and dressed in black when someone died, and their little boys aged and let their hair and beards grow wild . . .

Jack Farmer fell in love with the village at first sight. He found a room at a *domatia* above one of the cafés, dined on *souvlakia* and fried potatoes and a huge farmer's salad rich with black olives and salty *feta* cheese, then carried a bottle of oily yellow *retsina* down to the beach to sit crosslegged on the sand and watch the long arms of what Homer had called 'the wine-red sea' draw the flaming sunset down into its loving embrace . . .

It was early the next morning that he saw them.

He was back on the beach, armed for a day of ray-catching with suntan lotion, coco mat and dog-eared copy of Kazantsakis' *Report to Greco*.

A sudden noise from the cliff startled him, and he looked up, shading his eyes with his book.

High above him, only three levels from the top of the cliff, two figures stood on a ledge near one of the cave mouths. They were too far away and the sun was too bright for him to see them clearly: all he could make out was a woman with long brown hair and a brief dress of some dark material, and a bearded man, a head taller than his companion, naked except for a pair of shorts of the same dark fabric.

Leftover hippies, he chuckled to himself. *Somebody forgot to tell them the party's been over for years.*

But then the tall man raised his arm and Jack saw that he was holding a club in his fist and waving it menacingly. The woman screamed, a shrill, piercing cry which ripped through the stillness of the morning. The man swung his weapon with terrible force and she jumped back, trying desperately to avoid the murderous blow. She was too slow, though, and it caught her on the shoulder and flung her against the stone face of the cliff.

Jack stared wide-eyed at the scene unfolding above him, shocked motionless as the woman sank to her knees, sobbing. The man in dark shorts pointed angrily at the nearest cave opening, and she crawled brokenly towards it and disappeared into the blackness. The man watched her go, then shouldered his club and stormed off along the ledge, around a protruding boulder and out of sight.

Jack turned in a dazed circle and looked around him. He was the only person on the beach at that hour. No one else had seen what had happened. No one else had heard that poor woman's scream or witnessed the viciousness of the blow she had taken.

He found himself shaking with fury at the cold-blooded cruelty of the bearded hippie. 'Damn him!' he said aloud, spitting out the words as if they were soaked in bile. 'Damn him!'

He tossed his book to the sand and set off for the cliff at a run. The chain-link fence barely slowed him down: he hoisted himself over it and dropped to the other side in seconds. There was a narrow switchback path up to the first level of caves, and from there crude flights of stone steps connected each level with the one above it.

Half-way up the cliff he stumbled and fell, scraping his palms on the warm, jagged rocks. A dizzying distance below him, the waves rolled in sedately to wash the golden sands of the beach. There was movement in the village—an old woman pushing a cart towards the street of vegetable stalls, a café proprietor setting out tables and chairs on his wide terrace—but he was blind to the charm of the scene. He pushed himself away from the stone and continued up the face of the cliff.

He reached what he was sure was the level where he had seen them, and dashed recklessly along the narrow ledge; uncertain which of the caves she had crawled into, he peered eagerly into each of them as he came to it.

But except for the faded colours daubed on their walls, dying reminders of the long-ago tenancy of the flower children, each of the shadowy vaults he checked seemed empty.

He almost missed her. He saw nothing in the cave at first, but the faintest movement in the darkness caught his eye as he turned to go on, and he stuck his head inside again to be sure.

She was lying huddled at the rear of the cave, her back to him, legs drawn up protectively and thin arms hugging her bare and dusty knees. She was trembling with pain and fear, and an almost animal despair came from somewhere deep in her throat.

Jack approached her softly and laid a reassuring hand on her shoulder.

She jerked away from his touch with a cry, pressed herself tightly against the stone wall behind her. Unlike the other caves Jack had passed, the walls here were unpainted, barren.

He moved several paces backwards, out of the personal space he had invaded. 'It's all right,' he soothed her. 'I'm here to help you.'

At the sound of his voice, her eyes went wide; they shone vibrantly, catlike, in the dark. She said something Jack didn't catch, and stepped towards him hesitantly. He stood where he was and let her come to him.

As she neared, he found that he could see her more clearly. Her face was puffy with crying, her long brown hair hung limp and uncombed down her back, a vivid bruise purpled her shoulder where the club had caught her. She was young— twenty, perhaps, twenty-three or twenty-four at most—but her skin was older, baked rough and dry by the sun. She was not pretty, not at all, yet there was something about her, some strange magnetism which attracted Jack strongly.

Her lips were her best feature, perfectly bowed, full and moist and parted in wonder. In slow-motion, she raised a hand to his face and ran her fingers gently across his smoothly-shaven jaw. Her feline eyes glittered luminously, compellingly.

'My name is Jack,' he told her, swallowing away a sudden thirst. 'What's yours?'

She looked a question at him, touched a fingertip to his lips, did not answer.

'Do you speak English?' he asked. 'What's your name?' He reached into his memory for the bits and pieces of language he had picked up in his travels. '*To ono masou, parakalo?*' he tried, beginning with phrasebook Greek. '*Como te llama? Comment vous appelez-vous? Wie ist Ihr Name? Hoe heet jij?*'

She stared at him blankly, curiously, silent.

'Are you OK?' said Jack. 'Can you speak? Do you understand me?'

She looked down—shyly, he thought—and her gaze came to rest on his bathing trunks. She made a surprised noise and took the cloth between her fingers and felt it. When she looked up at him again, he saw confusion and awe written together on her face.

For the first time, he looked closely at the dress she was wearing. It was not made of fabric, as he had thought when he saw it from the beach. It was fur, coarse brown fur, skinned from some animal he had not yet spotted on the island. It was not stitched together, the garment, but simply draped around her body and tied with a length of vine.

'Who—who *are* you?' Jack asked her nervously, the words prickly in his throat. 'What are you doing here? Why don't you speak?'

And then his own eyes widened and he gaped, stunned, around the cave. 'The walls!' he cried. 'They're painted, everywhere else! Why aren't they painted in here?'

She smiled at him, and murmured something unintelligible.

He whirled about, frightened now, and rushed to the mouth of the cave. He looked out and down, saw the deserted beach and brooding sea—and froze.

Matala!

The village!

It was gone!

Where the sleepy hamlet had been—squares and street, low buildings and unhurried inhabitants—there was nothing to be seen but forest, dense, impenetrable greenery and the chatter of a million birds.

'Where am I?' he screamed. 'What *is* this place?'

And then the man in the dark shorts was there, not five feet away from him on the ledge. Incongruously, the first thing Jack noticed about him was that it was not a pair of shorts he was wearing, after all, but a coarse fur loincloth that divided his bronzed torso from his thick, muscular legs.

Only then did he register the two enormous fists clenched tightly about the butt end of the wooden club, and the savage fury etched coldly across the primitive countenance.

With a moan of fear, he flung himself backwards into the cave. The brute came after him, growling angrily and swinging his club with dreadful force. Jack backed away from the steady advance, staring in petrified fascination at the scything motion of the weapon, until he bumped up against the rear wall of the cave and could go no further.

'Stay away from me!' he shrieked, his mind drenched in panic. 'Stay away!'

There was an evil, animal grin on the bearded face, and the man drew back his club and came closer—

And, from nowhere, the strange young woman threw herself frantically at the creature's back, knocking him off-balance and sending his heavy club flying.

Jack jumped for it automatically, got his hands on it, and spun to face his attacker. The man in the loincloth had already scrambled to his feet, but he had forgotten about Jack for the moment. His powerful hands balled, his broad chest heaving, he was after the woman now, snarling with frustrated rage.

As he moved slowly towards her, Jack crept up behind him and hoisted the club high above his head. He brought it down with every ounce of strength in his body, felt it connect solidly and heard the creature's skull split open with one sickening crack.

Rippling muscles sagged, fisted fingers splayed wide, violent energy dissipated as if it had never been—and the tall man crumpled to the floor of the cave like a marionette whose strings had abruptly been severed.

Sobbing, screaming, Jack struck the fallen figure again and again, battered the broken head till all that remained was a shapeless mass of pulp and bone and blood.

At last he stopped, gasping hoarsely for breath. Hot tears stung his eyes as he saw what he had done. He felt the rough wood of the club chafing his raw, blistered hands, and threw it away from him with a cry of revulsion. He remembered the woman and how she had saved him, and turned around to look for her.

She wasn't there.

Must have run outside, he thought dazedly, and stumbled after her, half wanting to make sure that she was safe and half just to get away from the terrible thing on the cave floor.

Out on the ledge, she was nowhere to be seen.

But Matala was there, spread out peacefully below him, and a knot of figures were standing on the beach, staring up at him and pointing.

It never happened! Jack thought wildly. *It wasn't real!*

But when he spun around and peered back into the dimness of the cave behind him, the lifeless body that lay there on the floor was real.

And when he clutched his arms to his chest to stop the shivering, the wine-red blood that stained them was real.

And when he gazed down over the ledge to the beach and saw two uniformed policemen clambering over the chain-link fence and climbing rapidly towards him up the face of the cliff, he knew that they, too, were very, very real . . .

The Running - Companion

PHILIPPA PEARCE

Any day, over the great expanses of the Common, you can see runners. In track-suits or shorts and running-tops, they trot along the asphalted paths across the grass, or among the trees, or by the Ponds. On the whole, they avoid London Hill, towards the middle of the Common, because of its steepness. There is another reason. People climb the Hill for the magnificence of the view of London from the top; but runners consider it unlucky, especially at dusk. They say it is haunted by ghosts and horrors then. One ghost; one horror.

In his lifetime, Mr Kenneth Adamson was one of the daily runners. This was a good many years ago now. His story has been pieced together from what was reported in the newspapers, what was remembered by neighbours and eye-witnesses, and what may have been supposed to have been going on in the mind of Mr Adamson himself.

Sometimes Mr Adamson ran on the Common in the early morning; more often he ran in the evening after work. He worked in an office. He was not liked there: he was silent, secretive, severe. People were afraid of him.

The Adamsons lived in one of the terrace-houses bordering the Common. There was old Mrs Adamson, a widow, who hardly comes into this story at all; and her two sons, of whom Kenneth, or Ken, was the elder. There were only two people in the world who called Mr Adamson by his first name: they were his mother and his brother. He had no wife or girl-friend; no friends at all.

Mr Adamson ran daily in order to keep himself fit. The steady jog-trot of this kind of running soothed his whole being; even his mind was soothed. While his legs ran a familiar track, his mind ran along an equally familiar one. Ran, and then ran back, and then ran on again: his mind covered the same ground over and over again.

His mind ran on his hatred.

Mr Adamson's hatred was so well grown and in such constant training that at times it seemed to him like another living being. In his mind there were the three of them: himself; and his hatred; and his brother, the object of his hatred.

Of course, Mr Adamson's brother never ran. He could not walk properly without a crutch; he could only just manage to get upstairs and downstairs by himself in their own house. He had been crippled in early childhood, in an accident; and his mother had not only cared for him, but spoilt him. To Mr Adamson's way of thinking, she had neglected *him*. Jealousy had been the beginning of Mr Adamson's hatred, in childhood: as the jealousy grew, the hatred grew, like a poison tree in his mind. It grew all the more strongly because Mr Adamson had always kept quiet about it: he kept his hatred quiet inside his mind.

He grew up; and his hatred grew up with him.

For years now Mr Adamson's hatred had been with him, not only when he ran, but all day, and often at night, too. Sometimes in his dreams it seemed to him that his running-companion, his hatred, stood just behind him, or at his very elbow, a person. By turning his head he would be able to see that person. He knew that his hatred was full-grown now; and he longed to know what it looked like. Was it monster or man? Had it a heavy body, like his own, to labour uphill only with effort; or had it a real runner's physique, lean and leggy? He had only to turn his head and see; but in his dreams he was always prevented.

'Ken!'

His mother's thin old voice, calling his name up the stairs, would break into his dreams, summoning him down to breakfast. Mr Adamson breakfasted alone, listening to the sound of his brother moving about in his room above, or perhaps beginning his slow, careful descent of the stairs. Listening to that, it seemed to Mr Adamson that he heard something else: a friend's voice at his ear, whispering a promise: 'One day, Ken . . .'

One day, at last, Mrs Adamson died of old age. The two brothers were left alone together in the house on the edge of the Common. They would have to manage, people said. On the morning after the funeral, Mr Adamson prepared the day's meals, then went off to his office. At this time of year, he ran in the evenings, never in the mornings. It was the beginning of autumn and still pleasant on the Common in the evening, in spite of the mist.

Mr Adamson came home from work; and presumably the two brothers had supper, talked perhaps—although Mr Adamson never spoke to his brother if he could help it—and prepared for bed. Just before bedtime, as usual, Mr Adamson must have changed into his running shorts and top and training shoes and set off on his evening run.

Questioned afterwards, the neighbours said that the evening seemed no different from any other evening. But how were they to know? The Adamsons lived in a house whose party-walls let little noise through. Would they have heard a cry of fear: 'Ken—no!' Would they have heard a scream? The sound of a heavy body falling—falling—

Some time that evening Mr Adamson's brother fell downstairs, fatally, from the top of the stairs to the bottom. Whether he fell by his own mischance (but no, in all his life, he had never had an accident on those stairs), or whether he was pushed—nothing was ever officially admitted. But the evidence examined afterwards at least pointed to his already lying there at the foot of the stairs, huddled, still, when Mr Adamson went out for his evening run. Mr Adamson must have had to step over his dead body as he came downstairs, in his running gear, to go out on the Common.

It so happened that neighbours did see Mr Adamson leaving the house. He left it looking as usual—or almost as usual, they said. One neighbour remarked that Mr Adamson seemed to be smiling. He never smiled, normally. They saw no one come out of the house with him, of course. No one followed him.

Mr Adamson set off across the Common, as usual increasing his pace until it reached a jog-trot. This was the speed that suited him. Joints loosened; heartbeats and breathing steadied; the air was on his face; only the sky above him. His mind felt both satisfied and empty: free. This was going to be the run of his life.

He planned to run across the Common to the Ponds; then take the main exit route from the Common, leading to the bus terminus and shopping centre; but he would veer away just before reaching them, taking a side path that circled the base of London Hill; and so home.

When he got home, he would ring the doctor or the police, or both, to report his brother's accident. He had no fear of the police. No fear of anyone.

Now, as he ran, he began to get his second wind, and to feel that he could run for ever. No, the police would never catch up with him. No one could ever catch up with him.

Pleasantly he ran as far as the Ponds, whose shores were deserted even of ducks. Mist was rising from the water, as dusk descended from the sky. Mr Adamson wheeled round by the Ponds and took the path towards the terminus and shopping centre. He was running well; it seemed to him, superbly.

A runner going well is seldom aware of the sound of his own footfalls, even on an asphalted surface. But Mr Adamson began to notice an odd, distant echo of his own footsteps: perhaps, he thought, an effect of the mist, or of the nearness of London Hill.

Running, he listened to the echo. Unmistakably, running footsteps in the distance: a most curious effect.

Running, listening carefully, he began to change his mind. Those distant footsteps were neither his own nor an echo of his own, after all. Someone behind him was running in the same direction as himself, trotting so exactly at his own pace that he had been deceived into supposing echoes. The footsteps were not so very far in the distance, either. Although the pace was so exactly his own, yet the footsteps of the other runner seemed all the time to be coming a little nearer. The impossibility of this being so made Mr Adamson want to laugh, for the first time in many years. But you don't laugh as you run.

Very slightly Mr Adamson increased the pace of his running, and maintained it; and listened. The runner behind seemed also to have very slightly increased his pace: the footsteps were a little more rapid, surely, and clearer. Clearer? *Nearer*? Mr Adamson had intended to leave the main way across the Common only just before it reached the terminus and shops: now he decided to take a side path at once. It occurred to him that the runner might just be someone hurrying to catch a bus from the terminus. That supposition was a relief.

He turned along the side path; and the feet behind, in due time, turned too. They began to follow Mr Adamson along the side path, never losing ground, very slightly gaining it.

Mr Adamson quickened his pace yet again: he was now running rather faster than he liked. He decided to double back to the main path, across the grass.

The grass was soft and silent under his feet. He heard nothing of his own footfalls; he heard no footfalls behind him. Now he was on the main path again, and still could hear nothing behind him. Thankfully he prepared to slacken his pace.

Then he heard them. The runner behind him must have crossed the soundless grass at a different angle from his own. The strange runner's feet now struck the asphalt of the path behind Mr Adamson nearer than he could possibly have expected—much nearer.

The pace was still the same as his own, yet gained upon him very slightly all the time. He had no inclination to laugh now. He ran faster—faster. The sweat broke on him, ran into his eyes, almost blinding him.

He reached his intended turning off the main path and took it. The feet, in due time, followed him. Too late he wished that he had continued on the main path right to the bus terminus and the shops, to the bright lights of streets and buses and shops. But now he had turned back over the Common, duskier and mistier than ever. He had before him the long path winding round the base of London Hill before it took him home. It was a long way, and a lonely, unfrequented one at this time of evening. The Hill was straight ahead of him, and he knew there would be groups of people at the top, people who walked there in the evening to admire the view. Never before had he chosen to go where there were other human beings, just because they were other human beings, flesh and blood like himself. Now he did. He took the path that led directly to the summit of the Hill.

The evening strollers on the top of the Hill had been looking at the view, and one or two had begun to watch the runner on the slopes below. He was behaving oddly. They had watched him change course, and then double to and fro—'like a rabbit with something after it', as one watcher said.

'He's coming this way,' said another.

'Straight up the Hill,' observed someone else in the little crowd. Most of them were now peering down through the dusk. 'Straight up the Hill—you need to be young and really in training for that.'

Straight up the Hill he went, his heart hammering against his ribs, his breath tearing in and out of his throat, his whole body dripping with sweat. He ran and ran, and behind him came the feet, gaining on him.

On the Hill, they were all staring now at the runner. 'What's got into him?' someone asked. 'You might think all the devils in hell were after him.'

'He'll kill himself with running,' said a young woman. But she was wrong.

Now he was labouring heavily up the steepest part of the slope, almost exhausted. He hardly ran; rather, staggered. Behind him the feet kept their own pace; they did not slow, as his had done. They would catch up with him soon.

Very soon now.

He knew from the loudness of the following feet that the other runner was at his back. He had only to turn his head and he would see him face to face; but that he would not do—that he would never do, to save his very soul.

The footsteps were upon him; a voice close in his ear whispered softly—oh! so softly!—and lovingly—oh so lovingly! 'Ken!' it whispered, and would not be denied.

The watchers on the Hill peered down.

'Why has he stopped?'

'Why's he turning round?'

'What's he—Oh, my God!'

For Mr Adamson had turned, and seen what none of the watchers on the Hill could see, and he gave a shriek that carried far over the Common and lost itself in darkness and distance—a long, long shriek that will never be forgotten by any that heard it.

He fell where he stood, in a twisted heap.

When they reached him, he was dead. Overstrain of the heart, the doctor said later; but, being a wise man, he offered no explanation of the expression on Mr Adamson's face. There was horror there and—yes, something like dreadful recognition.

All this happened a good many years ago now; but runners on the Common still avoid London Hill, because of Mr Adamson and whatever came behind him. There may be some runners who fear on their own account—fear the footsteps that might follow *them*, fear to turn and see the face of their own dearest, worst wickedness. Let us hope not.

Gibson's

ANN PILLING

Our Aunt Mildred is the most practical person I know, and the least superstitious. She always walks under ladders, often wears green, and her city flat was number Thirteen. So the whole family was amused when she retired from teaching and bought a house on the end of a country village. The place she found had been up for sale for over a year, and she got it cheap, because nobody else would buy it.

'Well, I'm glad nobody *did*,' she said, showing us the estate agent's details. 'It's going to take every penny I've got as it is. Houses in those villages are always expensive. It's so convenient for Birmingham. Not that I'll want to go back there, once I've settled in.'

She was delighted. My little sister Josie looked at some photographs of the cottage while I read the details. It was built of stone and was very small: two rooms downstairs and two above, with a tiny box-room Aunt Mildred was planning to 'improve'. There was quite a bit of land with it. She had never had a garden before.

We could just make out a date, 1620, carved over the door. The cottage had an odd name, too: *Gibson's.*

'I expect you and David think it should be called something like *Honeysuckle Cottage*,' my aunt said, smiling at Josie. 'No, I like *Gibson's*. There's no nonsense about it. Now, when are you coming to stay?'

She didn't need to ask twice. We'd never been able to stay in her small flat, though she'd often visited us. She is actually our great-aunt, and my godmother, and very special. And she was as good as her word. Our phone rang before she'd been in her new house a week, and Aunt Mildred was soon arranging with my mother for Josie and me to stay with her during our October half-term holiday.

Gibson's was much better than the agent's description. Somehow, those typed leaflets never give you a sense of how a place *feels*. Perhaps it is because we have always lived in a modern house that I find old ones so fascinating, but the minute I walked into Aunt Mildred's I became very excited.

It smelt old. Every door, window, and floor-board creaked and rattled, and nothing was at right angles, or fitted properly. The beams in the living-room were black with age and spiked with rusty old hooks. There was even a well in the garden, and it worked. On our first night Aunt Mildred made a log fire in the old inglenook, and we sat around eating toast and staring into the flames.

'I'm never going back home, Auntie,' my sister said dreamily. 'I'm going to live here, with you.'

We all laughed. Josie was always saying things like that.

'Think of all those silly people who decided not to buy it,' Aunt Mildred said, buttering more toast. 'Lucky old me. They must have been mad.'

And I agreed with her; they must have been.

When we went outside after breakfast on our first morning, there was an old man digging near the well. He was called Joe Glover, and he was helping Aunt Mildred with the garden. It was a large plot, on three sides of the house, with grass and flowerbeds, and old fruit-trees straggling up a hill. Everything was very overgrown and weedy.

'It's been empty so long, y'see,' the old man said. 'Things've got a bit out of hand, like. But we'll soon sort it out. Nice old place this is—it just needs a bit of paint. You interested in old things, are you?'

We nodded. 'Well, you come down the village this afternoon. Vine Cottage I live in, next to the pub. I'll show you some old things, things I've dug up. You come this afternoon, after I've had me bit of dinner.'

In Joe Glover's shed was a big wooden box. He opened it and began to arrange the contents carefully on his work-bench. There were several large copper coins, as thick as three old pennies stuck together, two daggers without handles but with patterns delicately engraved along the blades, and a lump of black leather that looked as if it might once have been a shoe.

Just before he closed the box the old man reached down inside and drew something out very carefully. He blew the dust off, then put it on the bench very gently. It was a human skull.

'*Ugh!*' Josie said. 'Did you dig that up in Aunt Mildred's garden?'

'Oh, *no*,' he replied with a grin. 'There's nothing like that round *Gibson's*. No, I found these years ago, on the hill, when I was a boy.'

'Which hill?' asked Josie. I knew she was frightened. She didn't like the skull.

'Edgehill. There was a great battle there, hundreds of years ago. And people still find things now and then, in the fields, when the earth's been turned over: bits of armour and knife-heads, bones, too—the farmers, like, y'know. Most of it's in the museums now, but there's still finds.'

Of *course*, I was thinking, *Edgehill*. It was the first big battle of the Civil War. I remembered the crossed swords sign on the map, when we were trying to locate Aunt Mildred's cottage, and Dad explaining to Josie about the Cavaliers and the Roundheads. We had driven up the hill yesterday, in Aunt Mildred's old Mini, along a steep lane that wound round and up, through the trees, on to a long ridge which we'd seen miles back, from the flat fields below. She'd stopped at the top so we could get out and look at the view.

'Your auntie'll tell you all about King Charles and the Battle of Edgehill. She's a school mistress, isn't she? I expect she'll know a bit about it.'

She should do, I thought. She was a history teacher for nearly forty years.

Aunt Mildred's front door opened straight out on to a narrow road. If you turned right you walked down past the church and eventually reached the first cottages of the village. Left, the road climbed up a hill, petering out after a mile or so in a rough track. But if you carried on walking, along a kind of sunken lane, you came out on to Edgehill itself.

Aunt Mildred didn't waste time sitting down for meals, not when the sun was shining. 'Come on,' she said when we got back from Joe Glover's, 'let's take our tea with us and go for a walk.' We filled a bag with buns, apples and chocolate, and set off up the lane. The high, uncut hedges met over our heads, making a green, arched tunnel. Josie didn't like it much. It was too gloomy. She held on to Aunt Mildred's hand all the way.

From the top of Edgehill we could see for miles. The grassy slopes dropped sharply away under our feet, flattening out at last into a patchwork of fields which unfolded endlessly into the distance, becoming a green-gold blur, speared with the odd church tower. It was so quiet. There was only the buzz of a car on the main road down below, and the birdsong all around us, in the hedge.

'Hard to believe they all rushed together down there, hacking one another to death,' Aunt Mildred said, looking out over the lush Warwickshire plain. 'Some of them were only boys, too.'

'Who won?' I asked her. 'The king's men, or the Roundheads?'

'They both claimed victory. Some of the dead are buried in the villages round here.'

'Are there any in *your* village?' said Josie.

'I don't know. We must go and have a look.'

As we walked back we told Aunt Mildred about Joe Glover's collection. She laughed, 'I suppose he's told you all about his ghostly army, too? He bores the pants off everybody with that story.'

'No, he hasn't. What army?'

'Well, on the anniversary of the battle people claim to have seen figures moving about on Edgehill, shadowy horsemen and foot-soldiers, fighting on the plain down there. They've heard noises, too.'

'What sort of noises?'

'Oh, the sounds of battle. Swords clashing together, cannons, people getting their arms and legs chopped off,' she ended cheerfully.

'You don't really believe in it, do you, Auntie?' Josie said, round-eyed.

'Not a word. It's just a load of rubbish. But it makes a good story.'

After supper she found me sitting by the big bookcase on the landing.

'I know what you're looking for, David, and the answer's October 23rd, 1642. I read it all up the week I moved in.'

'That's the date of the battle, is it?'

She nodded.

After a minute I said, 'But, Aunt Mildred, that's this week! It's tomorrow, in fact.'

'Oh, I know *that*. And there's just one thing. If you decide to go ghost-spotting put an extra vest on. I don't want your mother after me if you get flu.'

But of course I didn't go. It was wet the next day, and by evening the drizzle had become a heavy downpour. I didn't fancy tramping up that sunken lane all on my own, in the dark. Anyway, I wasn't very interested in ghosts. As Aunt Mildred said, it was just a good story.

That night Josie woke up screaming. I opened my eyes and heard the clock downstairs strike two, then I heard her cry out. After the second cry her voice became a real shriek. By the time I got to the door of the box-room Aunt Mildred was there, sitting on the bed, with her arms round her.

'I can't *breathe*!' Josie was screaming. 'I just can't *breathe*! Let me get to the window. Give me some air!'

There was no window in the room she'd been sleeping in. It was just a box-room off the landing. She'd chosen it because it had pretty wallpaper and a fairy-tale latch on the door. Aunt Mildred was waiting for the builders to come and knock a window in. There was only a tiny sky-light, and that wasn't made to open.

She held Josie very tight. 'Just a nightmare,' she was saying. 'There's plenty of air, dear—look, the door's wide open. Now why don't you lie down again?' But Josie was still sobbing.

I switched the light on and looked at her. Her face was very white, and her hair was plastered to her head in damp curls. She was shaking violently.

'Let her stand by the landing window, Aunt Mildred. It's so stuffy in here,' I said.

In seconds she was calmer. The cold night air flowed in over us, and we all shivered slightly. Josie's eyes were closing; she was falling back against Aunt Mildred and starting to breathe quite heavily.

'She's asleep again,' my aunt said quietly. 'Do you think, between us, we could carry her into my bedroom? She can sleep with me tonight.'

Josie is quite small for eight. We lifted her quite easily and tucked her up in Aunt Mildred's bed. Five minutes later the house was quiet again. I whispered goodnight and shut my aunt's door. But I felt thoroughly awake now. What do you do, in the middle of the night, when you feel like running a hundred-metre sprint?

I could always read till I felt sleepy. I crept over to the landing bookcase, then noticed we'd left the window wide open. As I leaned out to shut it I glanced down into the garden. By the well there was a dark shape, standing out a foot or so from the old wall, in soft earth, where Joe Glover had pulled some bushes out the day before.

For a minute I thought it was a tree. Then it lifted its arms away from its sides. I wanted to shut the window, but my hand froze on the sill. A small wind sent leaves scurrying across the ragged grass, driving cloud off the moon's face, lighting up the garden. It plucked at the long cloak the figure was draped in, then moulded it round her.

I knew it was a woman because I heard her voice, a long, lamenting cry of great bitterness that echoed against the house. Surely I was dreaming? Ghosts do not cry out, nor do they communicate with those who see them. They live in another time from ours, in another world, and those worlds pass each other but never join.

I fastened the window and turned away from the garden, but the terrible crying went on, and as I looked out again, willing the woman to turn into an innocent tree, I saw that she was lifting up her arms to me and tossing her draped head from side to side, voicing that deep wild cry of utter desolation.

I lay in bed with the light on, all night, and when I fell at last into a broken, uneasy sleep I dreamed about Aunt Mildred charging down Edgehill on a great black horse, and about Joe Glover, trying to sell that skull to an estate agent in Banbury.

In the cool air of the morning it all seemed slightly ridiculous. Something had certainly disturbed Josie, but it was possibly no more than that awful stuffed-up feeling you get before a cold starts. She'd probably begin sneezing today. And I'd had peculiar dreams before, when I'd been sleeping away from home.

Even so, it was too vivid. If I didn't tell someone about it I would have it again. Aunt Mildred might say it was a lot of stuff and nonsense, but I knew she'd listen to me. I pulled my clothes on and clattered down the stairs, hoping to get her on her own, before Josie woke up.

She was in the kitchen, frying bacon. 'Aunt Mildred,' I began, 'I'm sorry about last night. I think Josie's getting a cold. But there's something else. When you'd gone back—'

Then the back door opened and my sister came in from the garden.

'Will David swap bedrooms with me?' she said at once. 'I'm not sleeping in that room again. It's too stuffy. I just couldn't get my breath.'

'What woke you up?' I asked her. 'You've never done that before. It's never happened at home.'

Josie screwed her mouth up and stared at the floor. I knew that look. It meant: 'Don't ask me any more because I'm not telling you.' She got it when things went wrong at school.

'You don't mind changing rooms, do you, David? Or you can sleep down here on the settee, if you'd rather,' Aunt Mildred said briskly, turning the bacon over. 'That little room will be so much better when the window's in. Which reminds me—I must chase up those builders. They're supposed to be coming on Thursday.'

After about three hours' sleep I felt the need of an early night, in a proper bed. 'No thanks, Aunt Mildred, I'll sleep in the box-room.'

I heard the downstairs clock strike two and turned over in bed. Something was skittering about in the roofspace, just over my head. Mice. Aunt Mildred was talking about getting a cat, now she lived in the country. She could certainly do with one.

I tried to get back to sleep. Then a door started banging somewhere. It was windy, and everything in the house was rattling. If I didn't go down and shut that door I'd be awake all night.

But there was something wrong with me. When I tried to raise myself from the mattress and get out of bed, I found I could not move, that I was unable to shift my position in any way. I just lay there helplessly under the blankets, willing my legs and arms to do something, trying to lift my head from the pillow. It was as if I had been turned to stone while I slept.

Then I felt pain, a deep stabbing in my back, and throbbing aches in my legs, pain so unbearable that a cry gathered in my throat. But I could not scream aloud, or make any kind of noise because something was pressing down on me, something enormously heavy that came down violently upon my chest, forcing the breath out of me, so hard it almost split my breast-bone, then covered my mouth so that the air came neither out nor in.

I was choking. If I didn't get out of that room and away from that crushing weight I would die. But my limbs were useless. It was as though all the central nerves were severed, and my legs and feet dangled down helplessly, as if I were some kind of stuffed doll.

In dreams, sometimes, come moments of blinding vision, and with them the strength of angels. Suddenly, something in the room snapped. I cannot say, now, whether it was a loud noise, or a flash of light. All I can remember is that whatever it was that bound me lost its strength, so that I leaped up screaming from the bed, with the feeling that a great stone had been rolled off me, that I had a chance of life.

I can remember collapsing on the landing, and leaning out of the window, gobbling in great lungfuls of the night air. My own voice, hysterical and high, was saying: 'I couldn't breathe . . . something was stopping me. It's just what happened to Josie.' And below there was an undertow, the low, passionate sobbing of a woman, mingling with my cries.

'Come along, David, get back to bed. This nonsense has got to stop, you know.' Aunt Mildred was there, leaning past me to shut the landing window. She snapped the catch down quickly, too quickly to see what I had seen, the same dark figure by the well, with its cloak streaming in the wind, lifting up despairing arms towards the house.

My aunt was in a bad mood next morning. 'There's no need to argue about who's sleeping in the box-room tonight,' she snapped at us. 'Because the builders are coming today to put the window in. What I *do* need is some help in clearing the room out for them.'

I was unable to get my sister on her own, to talk to her, because Aunt Mildred had us organized the minute breakfast was cleared away. Josie covered everything up with old sheets, and I helped my aunt dismantle the box-room bed and store it in a corner of her bedroom. Down in the kitchen she arranged mugs, biscuit tin, and electric kettle on a tray.

'There. That's ready. If I know them they'll wander upstairs and take a few measurements, get their tools out, then come straight down again and brew some tea. Come on, we're going out for the day. They might get on quicker if we're not here to talk to.'

It was a glorious wind-swept autumn morning. We wrapped up warmly and set off in the Mini, and spent the day rambling around the neighbouring hamlets of honey-coloured stone, throwing the remains of our sandwiches to village ducks, peeping into churches, listening to Aunt Mildred reading snippets of history out of her guide-book. She was quite cheerful again.

Nobody mentioned the box-room, or what had happened to me in the night. But I was determined to tell Josie. What I couldn't decide was what Aunt Mildred really thought about *Gibson's*, and whether she'd experienced anything odd there herself.

Joe Glover had puzzled me the day before, when he'd told me something very strange. First he'd dropped hints that the cottage was always changing hands, and that nobody stayed in it for very long. When I asked him why, he simply said: 'You ask your auntie. I reckon she knows.'

'But I'm sure she *doesn't*, Joe,' I answered. 'I'm sure she'd have told us.'

'Doesn't she, now?' he muttered, giving me a very direct stare with his blue eyes. 'I reckon that's why them builders are coming tomorrow, to knock a hole through.'

'What on earth do you mean?'

'There's something wrong somewhere, David. Always has been. Things'll be unhappy at *Gibson's* till the poor soul's had its release.'

'What "poor soul"?'

'I don't know. Don't suppose as anyone does. All I do know is that nothing'll be right till it's set free.'

'But what about the builders?'

There was a very long pause. The old man leaned on his spade and stared out across the village.

'I was born and bred here, David. I'm a Warwickshire man. But my grandmother was Irish. County Waterford she lived in, right down in the south. My mother took me to see her once, when I was a lad. Your age, I'd be.

'While we were there her own mother died, my great-grandmother, that was. That was why we'd gone, I suppose. It was a long journey in those days, from our village to hers, in Ireland. I was in the room when the old woman passed away. My grandmother got up from the bedside and opened the windows. "Why are you doing that, Grandma?" I said. "Hush, Joe," she told me. "The poor soul's gone. We must let her spirit leave this house now, and give it freedom, so it can go to its Maker."

'I never forgot that. Must be sixty years ago or more. That's what they always did when a person died, or there'd be unhappiness, y'see. I reckon that's why your auntie's getting them builders to knock that window in.'

'I don't think she knows anything about *that*, Joe,' I said, after a few moments. 'She would have told us.'

'Would she, now?' he said irritatingly. And those blue eyes of his threw doubt on everything.

Then something happened that reassured me about Aunt Mildred, and showed us she'd known nothing of *Gibson's*, or its history.

We'd come back to the cottage just after four o'clock, and she parked her Mini in the lane. When we got out Josie said: 'We've not been in *your* church, Auntie. Can we go now? Come on, you do live next door to it.'

My aunt looked at her watch. 'Well, all right, we'll have a quick look inside. But I mustn't miss the builders. They'll be going home soon.'

The tiny church was full of light. There was no stained glass, only small, plain windows, so sunshine flooded the chancel, and trees, waving outside, cast gently moving shadows. The low, plain roof was held up by chubby Norman columns, and on one wall we inspected the flaking remains of a great painting, with little stick people all trooping obediently into a dragon's mouth. 'The jaws of Hell,' Aunt Mildred explained knowledgeably.

At the back was a thick, faded curtain. There was nobody about so we lifted it up and went through. We were in a kind of small ante-room full of buckets and mops, with the vicar's robes hanging from a brass hook on the wall. The place was covered with memorial stones, like the rest of the church, but nobody had bothered to clean them up behind the curtain. The largest one covered half one wall, but we couldn't read it properly because somebody had propped a shelf full of hymn books against it.

Aunt Mildred was staring at the large piece of whitish stone, and I saw her eyes harden suddenly. 'Josie, David, help me move this bookcase out of the way, will you?'

'But, Aunt Mildred, won't the vicar—'

'*Do as I ask you.*'

Her voice was strange. I had never seen my practical, no-nonsense godmother look so agitated before. It took a few minutes to take the hymn books off the shelves and stack them neatly on the floor. But when we had pulled the bookcase away the whole of the stone tablet was exposed. The ancient, graceful lettering said this:

Near this stone lie buried the remains of THOMAS GIBSON, a poore boy of Oxford, who fought with the King's men in the battle of Edgehill and did receive mortal wounds.
Being carried to an house near the churche he did languish three days, and no physician could come by him for a grete tempest of wind. Therefore they tooke pillowes and smothered his face, so he did die speedily, this being the 27th day of October 1642, in the fifteenth year of his age.
Also SUSANNAH his mother, widow of WILLIAM GIBSON. She died in grete infirmitie of mind, 15th April 1643, in the fortieth year of her age.

'Gibson,' Josie said. 'That's the name of your house.'

Aunt Mildred did not speak. She just went on staring at the stone memorial and holding my sister's hand very tightly.

I saw it all, then: the boy being taken from the field of battle along the deep lane, kind hands bringing him inside the cottage, carrying him up to that airless room under the roof, trying vainly to ease his pain. Then a great storm blowing up and the doctor unable to get near with his crude remedies, and the youth crying out in agony.

As his mother Susannah must have cried, when she walked from Oxford, to be told her son was dead, smothered to death by those who watched over him—in mercy, because there was no hope.

Josie was good at reading but the strange language confused her.

'What does the last bit mean, Auntie? What does "in great infirmity of mind" mean? It's put in such a funny way.'

'Perhaps it means she went mad, when they told her that her son was dead. She was a widow. Perhaps he was all she had in the world. It was less than a year later that she died herself. Her heart was broken you see, dear. That's what it means, I think.'

She was speaking very quietly. Then I saw tears running down her face. I slipped my arm through hers.

'Come on, Aunt Mildred, let's put all these books back. Then we can go and see if the builders have brewed a pot of tea.'

Our last two days were pleasant ones. Nobody talked about ghosts and battles. Josie dreamed undisturbed in the bedroom overlooking the lane, and I slept on the settee; neither of us saw anything suspicious out in the garden. The builders crashed about cheerfully and finished the work on the window in record time. They left on Saturday morning, and we spent the afternoon trying to get the cottage straight again.

'Don't touch anything till tomorrow,' warned the foreman. 'Let everything dry and harden off, then you can open the window. Makes a grand little bedroom, doesn't it?'

It was quite different now. It would always be narrow and poky, but the window had made it remarkably light and lying in bed you could see into the garden and look across to the clutter of cottages behind the church. I decided to sleep there on my last night.

When I went to bed I dozed off almost at once. It was a deep, untroubled sleep, with no foul nightmares. In the early morning I woke up feeling happy and expectant, as if it was the first day of our holiday, and not the last.

When I looked at my watch I saw it wasn't quite five o'clock. I should have realized it was very early, from the light. The sun wasn't up yet, and the garden looked grey and damp. The smell of paint and putty was very strong, but I could open the window now. I sat up in bed and pressed the handle down gingerly, opening it just a crack.

As I pushed at the frame I felt something go past me, the lightest touch, just brushing my face, softer than fur—more like a leaf, or the mere passage of air. And out in the garden I saw, just for a second, that same dark figure standing by the well. But the arms were no longer stretched out, beseeching me, they were wrapped round in an embrace. It was as though a mother was bending down to comfort a little child.

I listened as the shape faded, afraid in case I heard again that terrible anguished voice. But the only noises were a dog barking down in the village and the first birds twittering. So I knelt on the bed and pushed the window right open, letting in the birdsong and the sweet morning air.

The Servant

ALISON PRINCE

Ginny ran down the path. Her mother shouted after her from the back door, 'When you've got a house of your own, my girl, you can make as much mess as you like. But you're not having your pocket money until you've tidied your bedroom!'

Ginny snatched her bike out of the shed, kicked the side gate open and set off down the lane. Her mother was waving her arms frantically and shouting something, but Ginny took no notice. Summer holidays were *awful*, she fumed, pedalling fast. Just because there was no school to go to, people treated you as if you were nothing at all—just a meek little figure who had to fit in with the rest of the household and not be noticed. A handy person to boss about. Run round to the shops, Ginny, dry the dishes, Ginny, tidy your room, Ginny. It was like being a *servant*.

Ginny came to the top of Bunkers Hill and let the bike freewheel down the long slope. The wind blew her hair back and made the hot morning cooler. Below her the green landscape spread out like a toy farmyard. Further down, Bunkers Hill crossed the busy main road and became Nebbutts Lane, leading through the distant fields to Cuckoo Wood where the bluebells grew so thickly in the spring. Much nearer, just before the cross-roads, a disused track wandered off to the right. Ginny touched her brakes to check the bike's speed as she approached the junction. Nothing happened.

Panic clutched at Ginny's heart like a cold hand. The sunny day was whistling past her with a speed which made her eyes run. She grabbed repeatedly at the useless brakes, remembering now that she had told her father when he came home from work last night that they needed adjusting. The brakes had been slack yesterday, but now they had completely gone. And she was hurtling towards the busy highway. To go out there at this speed meant almost certain death.

There was only one escape. The track. A milk float was coming up the hill towards her, threatening to block the entry to Ginny's haven unless she got there first. She crouched over the handlebars to increase the bike's breakneck speed and banked the bike hard to her right. She missed the oncoming milk float by inches and caught a glimpse of the driver's startled face as she shot down the stony, disused lane.

The bicycle jumped and rattled over the rough surface but, to Ginny's relief, the track began to level out as it narrowed to an overgrown path between dark trees and straggling banks of brambles. Impeded by the long grass, the bike slowed down and at last stopped. Ginny got off shakily. Her knees and elbows felt as if they had turned to water.

After a few minutes she bent down and looked at the bike's brakes. They had been disconnected and the blocks removed. Her father must have been intending to buy some new ones for her today. But why hadn't he *said*? True, she had been out at a disco last night, but he could have left a note or told her mother . . . Ginny had an uneasy memory of her mother shouting something after her as she rode off this morning, but she thrust the thought away. Her parents simply didn't *care*, she told herself with a new burst of anger after being so frightened.

And now what? She had come out with every intention of staying out until lunchtime and she didn't want to go crawling home again so soon, no doubt to be bossed about and scolded for not stopping to listen to what her mother had been saying. Ginny propped the bike against the ivy-clad trunk of a tree and stared round her. It was very, very quiet. The trees seemed almost to meet overhead, shutting out the sunshine. Ginny gave a little shiver. And then she heard the bell.

It was a faint, tinkling bell, very distant. It rang with a peremptory rapidness as if shaken by an impatient hand. Somebody wanted something, and quickly. Ginny pushed her hands into her jeans pockets and set off along the path, leaving the bike where it was. Since she had nothing to do, she might as well go and find out where the sound of the bell had come from.

The path went on between its high banks in such deep shade that it was almost like being in a tunnel. Daylight glowed at its far end as if promising a clearing, and Ginny walked towards it quickly. The bell rang again, sounding closer this time. Ginny emerged from the trees to find that she had come out further along the hillside. The path ended in a field of ripening barley. Butterflies danced in the sun.

On the sloping ground beside the field, slate-roofed behind a flint wall with a gate in the middle, stood a house. Heavy lace curtains were tied in loops at its windows, and its doorstep was spotlessly white. A plume of smoke ascended from its chimney straight into the windless sky. As Ginny stared at the house, the bell sounded again, a longer, rattling tinkle. A looped curtain twitched back in the ground-floor window to the left of the front door and a face looked out. White hair, a high-necked blouse and two black eyes which stared accusingly.

'Violet! Come along in at once!' snapped a dry voice, and a finger tapped on the pane.

Ginny glanced over her shoulder in case somebody called Violet was standing behind her, but she was alone. The butterflies danced above the motionless barley.

The bell tinkled again, and this time Ginny's hand reached for the latch of the gate and she found herself running up the path. The untrodden whiteness of the front doorstep warned her not to enter this way, and she darted round the side of the house to where blue-flowered periwinkles fringed a paved yard. The back door stood open.

The large, dim kitchen had a red tiled floor, and a huge wooden plate rack stood above the stone sink like an ominous, complicated cage. Ginny found that she was listening intently, in a kind of dread. She was waiting for the bell to ring. In a few moments its jangling tinkle sounded, so close that it was almost inside her head. She ran through the shadowed hall, where patches of red and blue light gleamed from a panel of stained glass in the front door, and tapped on the white-painted door to her right.

'Come *in*,' said the dry voice impatiently.

Ginny opened the sitting-room door. The fingers which gripped the small brass bell by its ebony handle were thin and bony, the hand blue-veined, veiled by a ruffle of lace from the tight silk sleeve. Tiny jet buttons ran up the narrow bodice to the cameo at the high neck, and then there was the white face, the mouth thin and pinched and the nose as craggy as a parrot's beak, the eyes astonishingly black under the elaborate pile of white hair.

'You are not to go outside, Violet,' said the woman. 'You belong in here, with me.'

Ginny found that she was standing with her hands behind her and her feet together, and almost smiled at her own sudden politeness. 'My name's Ginny,' she said.

'Not suitable,' said the woman heavily. The black eyes travelled slowly down Ginny's figure until they reached her plimsolls, then travelled up again. Vertical lines appeared above the lips as the mouth tightened a little more.

'Violet,' said the woman, 'you will wear your uniform at all times in this house, do you understand?'

'But I'm not—' began Ginny. Her voice petered out as the tight lips smiled grimly.

'Oh, yes, you are, my dear,' said the woman. 'My servants

have always been called Violet. So much more convenient. I am
Mrs Rackham, but you will call me madam, of course.'

Ginny shook her head in confusion. This could not be
happening. But she looked at the little brass bell with the ebony
handle, and stared into Mrs Rackham's black, unblinking eyes,
and knew that it was true.

'I have been *waiting* for my breakfast,' said Mrs Rackham.

Ginny stared guiltily into the black eyes, struggling to hold
on to the idea that Mrs Rackham's breakfast had nothing to do
with her, Ginny Thompson.

Mrs Rackham leaned forward a little. 'Light the spirit lamp,'
she instructed impatiently, 'then go to the kitchen and get my
breakfast.' The blue-veined hand gestured towards the table
which stood by the window, draped with a lace cloth over heavy
red chenille. On it stood eggshell-thin cups and saucers, a silver
teapot and sugar bowl and a thin-spouted brass kettle which,
supported on a brass stand, stood over a small burner. Ginny
moved towards it. At any rate, she thought, it was better than
hanging about at home. If she was treated like a servant there,
she might just as well play at being a servant here.

A box of matches lay beside the brass kettle. Ginny struck one and turned up the wick in its holder. It burned with a steady blue flame. The old woman was mad, of course, Ginny told herself. It wasn't unusual in old people. Her own granny had been very absent-minded, always calling Ginny by the name of a long-dead aunt, Flora, which was even worse than Violet.

'That's better,' said Mrs Rackham, darting a black-eyed glance at the spirit lamp. 'Now get along to the kitchen, quickly. When you bring my breakfast, you will be properly dressed.'

Ginny smiled and said, 'All right.'

Mrs Rackham looked outraged. 'That is not the way to answer,' she snapped. 'Say, "yes, madam." And curtsy.'

Ginny held out imaginary skirts and curtsied deeply as she had been taught at her ballet class.

Mrs Rackham seemed even more angry. 'Just a small bob, you stupid creature!' she hissed. 'Do you girls know *nothing* these days?'

Ginny expected to feel amused as she gave an obedient little bob, but as Mrs Rackham growled, 'That's better,' and the black eyes bored into Ginny's mind, the hidden smile shrivelled and died.

Ginny left the room with quick, neat footsteps, closing the door quietly behind her. As she made her way back to the kitchen the voice of reason in her mind urged her to walk out of the door and back along the lane to her brakeless bicycle, and start pushing it home. On the other hand . . . Mrs Rackham had to have her breakfast. Perhaps whoever looked after her had gone out for a while. No doubt they would be back.

The kitchen was cool and quiet. Greenish light filtered through a small, ivy-covered window, and a few flies circled aimlessly under the high ceiling. Gazing up at them, Ginny saw that a black dress and several white aprons hung from a wooden airer, and a starched white cap dangled from the end of one of its bars. She unhitched the airer's rope from its hook on the wall and released it hand over hand, lowering the airer so that she could reach the clothes. If she was going to humour the old lady's delusions, she might as well do the job properly.

But as Ginny peeled off her T-shirt and jeans she found that she was listening in a kind of terror for Mrs Rackham's bell; as if its demanding tinkle had dominated her whole life. She struggled into the black dress and did up the rows of buttons down the front and on each sleeve. She pulled on the thick black stockings which she also found on the airer, sliding up the pair of elastic garters which were looped round the airer's end beside the cap. Then she tied on a white, lace-edged apron and pulled the starched cap over her curly hair. She looked round for something more suitable than her plimsolls and found, neatly placed beside the wooden mangle, a pair of highly-polished black shoes, fastened by a single button.

The shoes fitted as if Ginny had always worn them. She pulled up the airer, then stared round the kitchen with increasing anxiety. What did madam have for breakfast? Plates of all sizes stood in the cage-like wooden plate rack, but there seemed to be no fridge and the pantry contained no muesli or cornflakes.

Mrs Rackham's bell rang.

Ginny jumped round, a hand to the high-buttoned neck of her dress. The voice of reason seemed to have deserted her, and she could only think that madam was waiting for her breakfast and that she, Violet, had failed to get it yet. She ran to the front room.

'Do I have to wait all day?' demanded Mrs Rackham. A vigorous spurt of steam was hissing from the brass kettle over its burner.

'I—I'm sorry, madam,' stammered Ginny. 'I didn't know what you wanted.'

'Two lightly boiled eggs, brown bread and butter cut in fingers, toast and marmalade,' said Mrs Rackham. 'Stupid girl. You can make the tea now you are here.'

Ginny went across to the table. She found an ornate tea caddy and put two spoonfuls of tea into the silver teapot. Then she picked up the brass kettle—and let it fall back into its stand with a gasp of pain. The handle was almost red hot. Tears sprang to Ginny's eyes as she nursed her stinging pain, but Mrs Rackham threw herself back in her chair, convulsed with cruel laughter. 'They all burn their hands!' she cackled delightedly. 'It's always the same—again and again!' Then, just as suddenly, she was angry. 'Turn the burner down, you idiot,' she snapped. 'The room is full of steam. And fetch a kettle holder.'

In the pale light of the kitchen, Ginny looked at her hand and saw the long red weal across the palm, and wanted to sit down and cry. But Mrs Rackham's bell was ringing, and she snatched the kettle holder from its hook beside the great black range and ran back to the sitting-room. She made the tea and said, 'I'll go and boil the eggs.'

'*When* you have moved the table within my reach,' said Mrs Rackham. 'And where is the milk?'

Ginny pushed the heavy table across to the old lady's chair, hampered by the stinging pain in her hand. Then she ran back to the kitchen for the milk, which she found by some kind of instinct in a small lidded churn on the pantry shelf. She snatched a blue jug from its hook and ladled some milk into it. The bell was ringing.

'That is a *kitchen* jug!' screamed Mrs Rackham as Ginny proffered the milk, and lace ruffles flew as a hand flashed out, sweeping the jug from Ginny's hand to smash against the sideboard. 'Clear all that mess up,' Mrs Rackham commanded, her face tight with fury, 'then bring my milk in the proper jug. Where are my eggs? Don't you dare boil them for more than three minutes!'

Ginny ran sobbing to the kitchen, found a small glass jug and filled it with milk then carried it back to Mrs Rackham, who said nothing. Milk dripped from the polished edge of the mahogany sideboard.

As Ginny went in search of a cloth she tried to recall the reasonable voice which told her that she did not belong here; but there was nothing in her mind except worry and guilt and the stinging of her burned hand. She found a rather smelly piece of rag and cleaned up the spilt milk as best she could, and picked up the pieces of the broken jug.

'Violet, *where* are my eggs?' enquired Mrs Rackham.

'Coming,' said Ginny desperately.

'Coming, *madam*!' shouted Mrs Rackham.

'Coming, madam,' Ginny repeated, and went out with a little curtsy, wiping her eyes on her sleeve. The bell tinkled and she turned back.

The black eyes were fixed upon her with a new energy as the tea was sipped, the cup returned with neat precision to its saucer. 'Have you done the fires?' asked Mrs Rackham. 'Black-leaded the grates, washed the hearths, swept the carpets, dusted? Cleaned the knives, whitened the doorstep, done the washing, scrubbed the kitchen floor? And what about the bedrooms? Are the beds clean and aired?'

'I don't know,' said Ginny helplessly. Tears overwhelmed her. 'I don't know, *madam*!' screamed Mrs Rackham.

Ginny fled to the safety of the kitchen, shaking. She found eggs in a large bowl, and an egg timer with red sand in the lower half of its double-bulged shape. She took a saucepan from the shelf, still crying a little, and filled it with hot water from the huge black kettle which steamed on the range.

Ginny found that her burned hand was beginning to blister. Like a remembered dream, a voice in her head told her that she did not have to stay here. There was a memory, too, of wearing different clothes. Trousers, a shirt made of soft stuff which left her arms bare . . . Ginny wiped her eyes on her black sleeve again, with a gesture so familiar that it seemed as if she had done it many times before. She gazed round the kitchen as if seeking those other garments, but the wooden chairs with a pattern of pierced holes in the seats were bare in the dim light, and the flies circled endlessly against the high ceiling.

The water in the saucepan began to bubble, and Ginny lowered in two eggs with a spoon, then turned over the egg timer. As the trickle of red sand began to run through the narrow neck, she got a brown loaf out of the earthenware bread crock and cut two slices, biting her lip because of the pain in her hand. Then she buttered the slices and cut them neatly into fingers.

When the eggs were done, she assembled a tray and carried it through the hall to Mrs Rackham. The sitting-room was dazzling after the dim kitchen, for sunlight poured in through the long window. Outside, the barley shimmered in the sun and the butterflies danced. Tears suddenly brimmed again in Ginny's eyes. She would never be free to walk through the fields, to come so fast down a steep hill that her eyes ran, but not with tears.

'Don't stand there gawping, Violet,' said Mrs Rackham. 'Put the things down here.'

Ginny obediently slid the tray on to the lace cloth.

'Where is my toast?' demanded Mrs Rackham.

'I—I didn't know how to make it,' Ginny faltered. Remotely, she remembered making toast by putting slices of bread into little slits in the top of—of what? She shook her head, confused. She had always been here. She would never leave. She would die here.

'With a toasting fork, you stupid girl, in front of the range,' said Mrs Rackham. She decapitated an egg then added, 'The spirit lamp has gone out. Light it.'

Ginny held a burning match to the wick, but no flame sprang up.

'Refill it,' snapped Mrs Rackham, waving an irritable hand towards the corner cupboard.

Ginny opened the tall, panelled door and took out the bottle of spirit. Violet, she thought as she gazed at its wonderful purple colour. Violet. Like me.

With the kettle holder she gingerly removed the top of the burner and filled up its reservoir with spirit. Her burned hand made her clumsy and the spirit spilled over and ran down on to the lace cloth, soaking through into the red chenille below it. Ginny shot a fearful glance at Mrs Rackham, but madam was probing an egg with the delicate silver spoon, and did not look up. Ginny fed the wick carefully back into the reservoir and fitted the top into place again. Once more she struck a match and applied it to the wick.

A blue flame leapt up, not only from the wick but from the whole top of the burner, following the spilt spirit down the brass stand and on the soaked cloth under it. Ginny shrieked with terror and jumped back, brushing against the uncorked bottle of spirit with her sleeve as she did so and knocking it over. More spirit gushed out, and sheets of flame sprang up, engulfing the kettle and its stand, the teapot, the cups, the table. Mrs Rackham began to scream, her mouth wide open in the white face, her blue-veined hands upraised. The red chenille cloth was ablaze, and the varnish on the heavy mahogany table legs was wrinkling as it caught fire. Flames began to leap up the side of the chintz arm chair where Mrs Rackham sat, still screaming. The skirt of her silk dress shrivelled as the flames licked across the chair, and Ginny saw that Mrs Rackham's legs were as twisted and useless as a rag doll's, encased in heavy contraptions of iron and leather.

Outside, the dancing butterflies shivered behind a screen of heat as the looped curtains burned. The room filled with smoke, and Ginny began to gasp for breath. Suddenly she realized that she must get out. She could not help Mrs Rackham. The hem of her long black dress was beginning to smoulder as she ran from the room. She grappled with the bolt on the front door. Her dress was burning. The house was full of fire, and the red and blue stained glass windows in the front door were dimmed with the choking smoke.

As Ginny wrenched the door open and daylight burst upon her like an explosion it seemed that Mrs Rackham was screaming a single word, senselessly and repeatedly.

'Again!' she shrieked, and it was like a mad song of agony and triumph. 'Again! Again! Again!' And Ginny knew what the terrible word meant. Like a recurring nightmare whose end only leads to the next beginning, she was condemned to repeat this experience over and over again. Even now, as the air fanned her burning dress into greedy flames and the screams were swallowed up in the inferno which had been a house; even in the agony of burning alive, Ginny was listening for the tinkle of Mrs Rackham's bell. It would all begin again.

Somebody was shaking her. 'Ginny!' a voice was saying urgently. 'Are you all right? What are you doing here?'

Mrs Thompson stared down at her daughter, who lay huddled by the rusted gate in the flint wall, an arm flung protectively across her face. She appeared to be asleep.

'Again,' said Ginny, and trembled.

'Are you all right?' Mrs Thompson repeated. 'I was frantic when you went off like that—your dad said to tell you about the bike, that he'd get new brake blocks. He'll murder me. Then the milkman said you nearly crashed into him tearing down Bunkers Hill—well, I got the car out straight away and came down here looking for you.'

Ginny's eyes were open but she was not seeing her mother. Her gaze searched the sky with a kind of despair. 'Butterflies,' she murmured. Tears welled up and she rubbed her eyes on the back of her wrist wearily.

'Darling, don't cry,' said her mother. 'It's all right—I'm not cross or anything. I mean, it was partly my fault.' After a pause she went on, 'I found your bike along the lane. But why did you come here? I hate ruined houses, they're so creepy.' Rose bay willow herb, the fire weed, stood tall among the blackened heaps of stone. It really was a horrible place, Mrs Thompson thought. Some distance away the remains of a brass kettle lay dented and squashed in the sun.

Ginny stood up and brushed fussily at her bare arms, fiddling at her wrists as though buttoning tight cuffs. Her mother watched with dawning concern as the girl straightened an apron, smoothed out a long skirt, her anxious hands not touching the surface of her jeans. 'I must go,' she said.

'You're not going anywhere,' said Ginny's mother. 'You're coming home with me. You must have had a nasty shock. We can put your bike in the back of the car.'

Ginny gave a sudden start. 'I must go,' she said again with worried alertness. 'Mrs Rackham wants her breakfast. That's her bell. What am I doing out here?'

Her mother stared. 'Mrs Rackham? This house is known as Rackham's, yes, but there's nobody here now. Some old crippled woman owned it, they say, but she died in the fire when it was burnt down, along with some poor little servant girl.'

'Violet,' agreed Ginny. She dropped a small curtsy, not looking at her mother, and called, 'Coming, madam!' Then she set off with oddly neat little footsteps through the weed-grown rubble, trotting parallel to the garden wall until she turned at a right angle and ran on to where some blue-flowered periwinkle bloomed among the stone. Her mother intercepted her and caught the girl by the hand. Ginny flinched violently. A long, red, blistered weal lay across her palm.

'How on earth did you do that?' demanded her mother. 'There's nothing hot on a bicycle. Unless—you didn't put your hand on the tyre, did you, to try to stop?'

But Ginny did not hear. She was staring into the black eyes again, watching the thin mouth in the white face as the orders were snapped out, hearing the cruel laughter as she burned her hand again. Outside the tall window, the barley shimmered in the summer sun and the butterflies danced. But Ginny would never be free to walk among them again. She was Mrs Rackham's servant and madam wanted her breakfast. Again— and again—and again.

The Ring

and the

Rib

J. J. RENEAUX

It was a stormy, grey Saturday morning when Irene, the widow woman, went to clean the church. Irene was a poor woman and couldn't afford to put but a few *sous* into the collection basket on Sundays. Still, she did what she could. She cleaned and polished the church floors until the old wood glowed. She brought deep red roses and creamy gardenias from her own bushes for the altar, and their sweet scent mixed with the warm smell of wax and candles. Her work turned the plain little church into a place of beauty and hope.

The building was made of whitewashed boards and built shotgun-style. No fancy stained glass here, only square panes of blue-coloured glass in the windows. The sanctuary was as simple and unadorned as the poor people of the village themselves: old handmade pews worn smooth with age; an embroidered altar cloth; and the statues of the saints, once brightly painted, now fading with age.

On that morning, Irene lit the candles and a lamp, for the storm made the little church darker than usual. As she worked in the stillness of the room, a clap of thunder in the distance caused the board-frame church to tremble, and a shiver ran down her back. She had the uneasy feeling that she was not alone, that something or someone was hiding in the shadows. She listened but heard nothing.

'Just the storm comin' up,' she thinks. 'This ol' church is always full of squeaks and creaks when there's a strong wind blowin'.'

She worked on but could not get over the feeling that somebody was looking over her shoulder. Yet as often as she glanced about, she saw nothing. Then all of a sudden, a cold breeze touched her neck like the grip of an icy hand. She spun around to see a figure step out of the shadows.

It was a tall, bony man with a pale face and eyes that seemed to burn right through her. He wore high boots caked with mud, and his clothing was torn and dirty. In his side, from a bloody, gaping wound, a single bony rib glinted in the candlelight.

'*Mon Dieu!*' she whispers, clutching her rosary beads. 'Help me, help me!' Her heart throbbed in her throat as the ghostly man slowly approached.

He stretches out his bony pale hands and moans, '*Sauvez mon âme*. Help me! My soul can not rest. Please help me!' Tears flowed from his burning eyes and fell on the polished wood floor.

Irene was frozen with fear, but at last she found her voice and whispers, 'What do you want from me?'

'I have been murdered! I cannot rest in peace until justice is done. Find my murderer. Bury me.'

'Spirit, who killed you?' the woman asks.

'It was Villien,' the spirit moans. 'Villien murdered me. Find him! Lay my bones to rest. I am so weary, so tired. Help me!'

'Villien,' she whispers, 'he's the richest man in the parish. He owns everything. Nobody will believe a ghost told me this. They'll say I'm crazy. Villien will run me out of the parish! I've got children. We'll have no place to go. We'll starve. No, spirit, I can't help you, I'm afraid.'

But the ghost says, 'Don't be afraid. I will help you. Hold out your hand.'

Irene obeyed, and the spirit dropped something shiny in her palm: a golden ring. He held out his own pale hand and whispers, 'Give me your gold wedding ring.'

The woman took the ring from her trembling hand and offered it. The ghost moans, 'You must help me. Find Villien. Bury me!'

'Spirit, where are your bones hidden?' she asks.

'I will reveal the place when the time is right. Go now!' And with that, the ghost disappeared.

Irene waited in the dark church for a long time while the thunder continued to pound and blue lightning flashed through the windows. Was she dreaming? But the golden ring glowed in

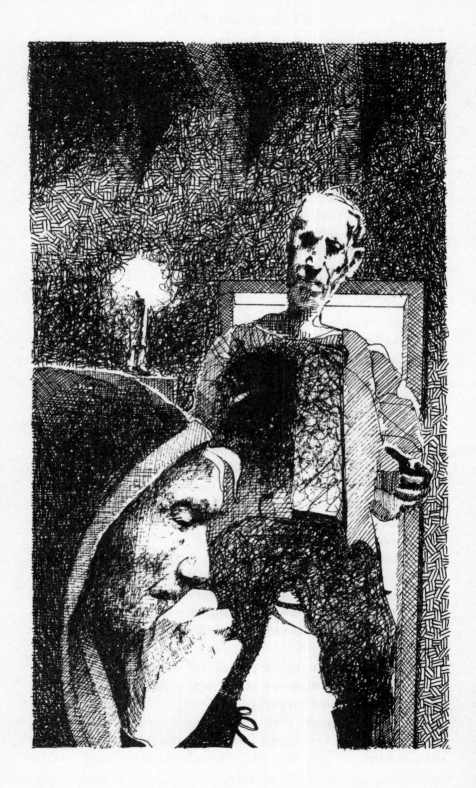

her hand, and on the floor the spirit's teardrops sparkled like dewdrops in the candlelight.

Finally the storm rumbled away and a gentle rain began to fall. Irene hurried to the village. She stopped before the office of *le grand shérif*. Her heart still beat hard, but the golden ring clenched in her fist gave her confidence as she stepped inside to tell her strange story.

'Shérif,' she says, 'a man has been murdered. You gotta come right away!'

'What's this? A man murdered?' he asks, jumping up from his chair. 'What man? Where is the body?'

'I don't know what man,' Irene says, shaking her head, 'and I don't know where the body is hidden.'

The sheriff looks at Irene in amazement. 'What do you mean, you don't know the murdered man and you don't know where the body is? How do you know there's been a murder?'

Irene looked him straight in the eyes and says, 'I know a man was murdered because his ghost appeared before me in the church and told me so. The ghost said that Villien was the murderer.'

'What? Woman have you gone moon-mad? Villien? You must think I'm crazy, too!' the sheriff hollers. 'You think I'm gonna arrest Villien, the richest man in the parish, with no proof, no evidence, no body?'

'The spirit gave me something to show you.'

Irene held out the golden ring. The sheriff took the ring, turning it round and round. Three initials had been carved inside the ring, the initials of a man who had disappeared five years ago.

'You say you found this ring at the church?' he asks.

'No,' says Irene, 'I said the ghost of a murdered man gave it to me, and he took my wedding ring. He begged for help. Said, "Find my body. Bury me." Said Villien killed him.'

'Something strange is goin' on for true out at that church,' the sheriff says, 'and I mean to get to the bottom of it!'

The *grand shérif* rounded up a group of men. They gathered up shovels and buckets and headed for the church. By the time they reached the little shotgun building, the rain had let up. The men began to dig all around the church, but their shovels didn't turn up a thing.

Back in the village, the story spread like wildfire until it reached the ears of Villien himself. His blood boiling, he saddled his horse and rode like the wind to the church.

Villien pulled his horse to a stop and glared at Irene.

'You accuse me of murder? Where is the body? You have no proof!' he hisses. 'Woman, you be outta my parish by sundown, you and your brats. I don't care where you go. You can starve for all I care! Nobody accuses Villien of murder and stays in my parish. Nobody will help you. They'll do as I say. They all work for me. I own them and their pitiful little village.'

The men all stopped digging, for they were afraid of Villien.

'There's nothing buried out here,' they cried. 'This woman is mad. Nobody's been murdered.' And they commenced to load up their tools.

Just then a voice whispered in Irene's ear, 'Look under the church. Dig under the church.'

'Dig under the church!' she calls out. 'Quick! Look under the church. The spirit told me.'

The men looked at each other and grumbled, but the sheriff ordered them to start digging under the church as the woman said.

Villien starts screaming, 'Stop that diggin'! Sheriff, if you wanta keep that badge, tell your men to stop.'

But the sheriff only hollered out, 'Keep diggin'!'

The little church was built up on rock pilings. Underneath, the air was musty and the ground was muddy. The men dug as best they could in the tight space. Suddenly, one of them gave out a shout that his shovel had struck something hard under the altar. Lifting his shovel into the light, he found a skull staring back at him.

Everyone was shouting with excitement and disbelief as they pulled out the remainder of the skeleton from its shallow grave. But a sudden quiet fell over them.

'There!' someone said, pointing.

'What is it?'

Something was shining through the dirt that clung to the skeleton. The sheriff carefully brushed the earth away. There, just as Irene had said, was her golden wedding ring, twisted tightly on to one bony rib.

The sheriff ordered the arrest of Villien. Terrified by the dead man's ghost, Villien confessed in full and found his justice at the end of a rope.

The villagers buried the skeleton in the cemetery, but it is said that the ghost continues to haunt the church. On Saturday mornings, women still come to wax and polish and set out roses and gardenias to glorify the altar. Now and then a cold shiver is felt, and they peer into the shadows and retell the story of the rib and the ring. Are they afraid? *Mais non.* It is a good spirit that watches over the poor little church, protecting all who come in need or trouble. Those who enter in peace have nothing to fear.

A Pair
of
Muddy Shoes

LENNOX ROBINSON

I am going to try to write it down quite simply, just as it happened. I shall try not to exaggerate anything.

I am twenty-two years old, my parents are dead, I have no brothers or sisters; the only near relation I have is Aunt Margaret, my father's sister. She is unmarried and lives alone in a little house in the country in the west of county Cork. She is kind to me and I often spend my holidays with her, for I am poor and have few friends.

I am a school-teacher—that is to say, I teach drawing and singing. I am a visiting teacher at two or three schools in Dublin. I make a fair income, enough for a single woman to live comfortably on, but father left debts behind him, and until these are paid off I have to live very simply. I suppose I ought to eat more and eat better food. People sometimes think I am nervous and highly strung: I look rather fragile and delicate, but really I am not. I have slender hands, with pale, tapering fingers—the sort of hands people call 'artistic'.

I hoped very much that my aunt would invite me to spend Christmas with her. I happened to have very little money; I had paid off a big debt of poor father's, and that left me very short, and I felt rather weak and ill. I didn't quite know how I'd get through the holidays unless I went down to my aunt's. However, ten days before Christmas the invitation came. You may be sure I accepted it gratefully, and when my last school broke up on the 20th I packed my trunk, gathered up the old sentimental songs Aunt Margaret likes best, and set off for Rosspatrick.

It rains a great deal in West Cork in the winter: it was raining when Aunt Margaret met me at the station. 'It's been a terrible month, Peggy,' she said, as she turned the pony's head into the long road that runs for four muddy miles from the station to Rosspatrick. 'I think it's rained every day for the last six weeks. And the storms! We lost a chimney two days ago: it came through the roof, and let the rain into the ceiling of the spare bedroom. I've had to make you up a bed in the lumber-room till Jeremiah Driscoll can be got to mend the roof.'

I assured her that any place would do me; all I wanted was her society and a quiet time.

'I can guarantee you those,' she said. 'Indeed, you look tired out: you look as if you were just after a bad illness or just before one. That teaching is killing you.'

That lumber-room was really very comfortable. It was a large room with two big windows; it was on the ground floor, and Aunt Margaret had never used it as a bedroom because people are often afraid of sleeping on the ground floor.

We stayed up very late talking over the fire. Aunt Margaret came with me to my bedroom; she stayed there for a long time, fussing about the room, hoping I'd be comfortable, pulling about the furniture, looking at the bedclothes.

At last I began to laugh at her. 'Why shouldn't I be comfortable? Think of my horrid little bedroom in Brunswick Street! What's wrong with this room?'

'Nothing—oh, nothing,' she said rather hurriedly, and kissed me and left me.

I slept very well. I never opened my eyes till the maid called me, and then after she had left me I dozed off again. I had a ridiculous dream. I dreamed I was interviewing a rich old lady: she offered me a thousand a year and comfortable rooms to live in. My only duty was to keep her clothes from moths; she had quantities of beautiful, costly clothes, and she seemed to have a terror of them being eaten by moths. I accepted her offer at once. I remember saying to her gaily, 'The work will be no trouble to me, I like killing moths.'

It was strange I should say that, because I really don't like killing moths—I hate killing anything. But my dream was easily explained, for when I woke a second later (as it seemed), I was holding a dead moth between my finger and thumb. It disgusted me just a little bit—that dead moth pressed between my fingers, but I dropped it quickly, jumped up, and dressed myself.

Aunt Margaret was in the dining-room, and full of profuse and anxious enquiries about the night I had spent. I soon relieved her anxieties, and we laughed together over my dream and the new position I was going to fill. It was very wet all day and I didn't stir out of the house. I sang a great many songs, I began a pencil-drawing of my aunt—a thing I had been meaning to make for years—but I didn't feel well, I felt headachy and nervous—just from being in the house all day, I suppose. I felt the greatest disinclination to go to bed. I felt afraid, I don't know of what.

Of course I didn't say a word of this to Aunt Margaret.

That night the moment I fell asleep I began to dream. I thought I was looking down at myself from a great height. I saw myself in my nightdress crouching in a corner of the bedroom. I remember wondering why I was crouching there, and I came nearer and looked at myself again, and then I saw that it was not myself that crouched there—it was a large white cat, it was watching a mouse-hole. I was relieved and I turned away. As I did so I heard the cat spring. I started round. It had a mouse between its paws, and looked up at me, growling as a cat does. Its face was like a woman's face—was like my face. Probably that doesn't sound at all horrible to you, but it happens that I have a deadly fear of mice. The idea of holding one between my hands, of putting my mouth to one, of—oh, I can't bear even to write it.

I think I woke screaming. I know when I came to myself I had jumped out of bed and was standing on the floor. I lit the candle and searched the room. In one corner were some boxes and trunks; there might have been a mouse-hole behind them, but I hadn't the courage to pull them out and look. I kept my candle lighted and stayed awake all night.

The next day was fine and frosty. I went for a long walk in the morning and for another in the afternoon. When bedtime came I was very tired and sleepy. I went to sleep at once and slept dreamlessly all night.

It was the next day that I noticed my hands getting queer. 'Queer' perhaps isn't the right word, for, of course, cold does roughen and coarsen the skin, and the weather was frosty enough to account for that. But it wasn't only that the skin was rough, the whole hand looked larger, stronger, not like my own hand. How ridiculous this sounds, but the whole story is ridiculous.

I remember once, when I was a child at school, putting on another girl's boots by mistake one day. I had to go about till evening in them, and I was perfectly miserable. I could not stop myself from looking at my feet, and they seemed to me to be the feet of another person. That sickened me, I don't know why. I felt a little like that now when I looked at my hands. Aunt Margaret noticed how rough and swollen they were, and she gave me cold cream which I rubbed on them before I went to bed.

I lay awake for a long time. I was thinking of my hands. I didn't seem to be able not to think of them. They seemed to grow bigger and bigger in the darkness; they seemed monstrous hands, the hands of some horrible ape, they seemed to fill the whole room. Of course if I had struck a match and lit the candle I'd have calmed myself in a minute, but, frankly, I hadn't the courage. When I touched one hand with the other it seemed rough and hairy, like a man's.

At last I fell asleep. I dreamed that I got out of bed and opened the window. For several minutes I stood looking out. It was bright moonlight and bitterly cold. I felt a great desire to go for a walk. I dreamed that I dressed myself quickly, put on my slippers, and stepped out of the window. The frosty grass crunched under my feet. I walked, it seemed for miles, along a road I never remember being on before. It led uphill; I met no one as I walked.

Presently I reached the crest of the hill, and beside the road, in the middle of a bare field, stood a large house. It was a gaunt three-storeyed building, there was an air of decay about it. Maybe it had once been a gentleman's place, and was now occupied by a herd. There are many places like that in Ireland. In a window of the highest storey there was a light. I decided I would go to the house and ask the way home. A gate closed the grass-grown avenue from the road; it was fastened and I could not open it, so I climbed it. It was a high gate but I climbed it easily, and I remember thinking in my dream, 'If this wasn't a dream I could never climb it so easily.'

I knocked at the door, and after I had knocked again the window of the room in which the light shone was opened, and a voice said, 'Who's there? What do you want?'

It came from a middle-aged woman with a pale face and dirty strands of grey hair hanging about her shoulders.

I said, 'Come down and speak to me; I want to know the way back to Rosspatrick.'

I had to speak two or three times to her, but at last she came down and opened the door mistrustfully. She only opened it a few inches and barred my way. I asked her the road home, and she gave me directions in a nervous, startled way.

Then I dreamed that I said, 'Let me in to warm myself.'

'It's late; you should be going home.'

But I laughed, and suddenly pushed at the door with my foot and slipped past her.

I remember she said, 'My God,' in a helpless, terrified way. It was strange that she should be frightened, and I, a young girl all alone in a strange house with a strange woman, miles from anyone I knew, should not be frightened at all. As I sat warming myself by the fire while she boiled the kettle (for I had asked for tea), and watching her timid, terrified movements, the queerness of the position struck me, and I said, laughing, 'You seem afraid of me.'

'Not at all, miss,' she replied, in a voice which almost trembled.

'You needn't be, there's not the least occasion for it,' I said, and I laid my hand on her arm.

She looked down at it as it lay there, and said again, 'Oh, my God,' and staggered back against the range.

And so for half a minute we remained. Her eyes were fixed on my hand which lay on my lap; it seemed she could never take them off it.

'What is it?' I said.

'You've the face of a girl,' she whispered, 'and—God help me—the hands of a man.'

I looked down at my hands. They were large, strong and sinewy, covered with coarse red hairs. Strange to say they no longer disgusted me: I was proud of them—proud of their strength, the power that lay in them.

'Why should they make you afraid?' I asked. 'They are fine hands. Strong hands.'

But she only went on staring at them in a hopeless, frozen way.

'Have you ever seen such strong hands before?' I smiled at her.

'They're—they're Ned's hands,' she said at last, speaking in a whisper.

She put her own hand to her throat as if she were choking,

and the fastening of her blouse gave way. It fell open. She had a long throat; it was moving as if she were finding it difficult to swallow. I wondered whether my hands would go round it.

Suddenly I knew they would, and I knew why my hands were large and sinewy, I knew why power had been given to them. I got up and caught her by the throat. She struggled so feebly; slipped down, striking her head against the range; slipped down on to the red-tiled floor and lay quite still, but her throat still moved under my hand and I never loosened my grasp.

And presently, kneeling over her, I lifted her head and bumped it gently against the flags of the floor. I did this again and again; lifting it higher, and striking it harder and harder, until it was crushed in like an egg, and she lay still. She was choked and dead.

And I left her lying there and ran from the house, and as I stepped on to the road I felt rain in my face. The thaw had come.

When I woke it was morning. Little by little my dream came back and filled me with horror. I looked at my hands. They were so tender and pale and feeble. I lifted them to my mouth and kissed them.

But when Mary called me half an hour later she broke into a long, excited story of a woman who had been murdered the night before, how the postman had found the door open and the dead body. 'And sure, miss, it was here she used to live long ago; she was near murdered once, by her husband, in this very room; he tried to choke her, she was half killed—that's why the mistress made it a lumber-room. They put him in the asylum afterwards; a month ago he died there I heard.'

My mother was Scottish, and claimed she had the gift of prevision. It was evident she had bequeathed it to me. I was enormously excited. I sat up in bed and told Mary my dream.

She was not very interested, people seldom are in other people's dreams. Besides, she wanted, I suppose, to tell her news to Aunt Margaret. She hurried away. I lay in bed and thought it all over. I almost laughed, it was so strange and fantastic.

But when I got out of bed I stumbled over something. It was a little muddy shoe. At first I hardly recognized it, then I saw it was one of a pair of evening shoes I had, the other shoe lay near it. They were a pretty little pair of dark blue satin shoes, they were a present to me from a girl I loved very much, she had given them to me only a week ago.

Last night they had been so fresh and new and smart. Now they were scratched, the satin cut, and they were covered with mud. Someone had walked miles in them.

And I remembered in my dream how I had searched for my shoes and put them on.

Sitting on the bed, feeling suddenly sick and dizzy, holding the muddy shoes in my hand, I had in a blinding instant a vision of a red-haired man who lay in this room night after night for years, hating a sleeping white-faced woman who lay beside him, longing for strength and courage to choke her. I saw him come back, years afterwards—freed by death—to this room; saw him seize on a feeble girl too weak to resist him; saw him try her, strengthen her hands, and at last—through her—accomplish his unfinished deed . . . The vision passed all in a flash as it had come. I pulled myself together. 'That is nonsense, impossible,' I told myself. 'The murderer will be found before evening.'

But in my hand I still held the muddy shoes. I seem to be holding them ever since.

The Hard-Working Ghost

DAL STIVENS

Misery-guts Jackson, the meanest cocky in the west, was going to bed just after sundown to save lighting a candle when he looked up and saw a ghost. Misery pulled the chaff-bag bedding over his head and stayed under for five minutes. When he peeped out the ghost was still there, leaning nonchalantly against the doorpost of the tin hut. He was about six feet high, well built for a ghost, and greeny-white, even to his dungarees, opened waistcoat, flannel shirt and whiskers. He looked about thirty-five but the colour might have aged him.

'W-hat d-do y-you w-want?' asked Misery.

'A job,' said the ghost. 'They tell me you're a good boss . . .'

'Good as they come,' said Misery, 'but I've no money to waste. Any moment now I'll be ruined.'

'I don't want any wages,' said the ghost. 'I'll work for nothing so long as I've got something to do. Haunting isn't enough to keep a big healthy bloke like me occupied. I'm desperate for a job.'

Misery sat up then.

'I suppose you'd want something to eat even though you're a spook?' he asked.

'Don't eat a thing,' said the ghost. He looked about forty-four round the chest.

'Can you drive a team?' asked Misery.

'With the best of them,' said the ghost. He flicked an imaginary whip.

'Can you chop wood and clear scrub?'

'None better,' said the ghost and swung his arms.

'Shear sheep?'

'On my head,' said the ghost and moved his thumb and forefinger together smartly, though no sound came forth.

'Mend harness?'

'Lead me to it!'

'Break horses?'

'No trouble,' said the ghost.

Misery put both legs over the stretcher.

'Got a match?' he asked. 'I can't see you very well.' The ghost struck a match and held it up. 'Things are tough,' said Misery. 'I couldn't give you any tobacco.'

'Don't smoke,' said the ghost. 'All I want is a job.'

Misery stood up. A gust of wind came in the door and the ghost wavered like a candle-flame.

'You don't look as though you could do a good day's work,' said Misery, sitting down on the stretcher.

The ghost threw away the match, lit another, flexed his arm muscles and then thumped his chest, which gave out a hollow sound.

'You look a bit light on it,' said Misery.

'I can make myself solid, if you'll only give me a job,' said the ghost. He knitted his brows as though concentrating, and when he thumped his chest again, the sound was sonorous and heavy. Misery stood up, walked over to the ghost and began punching him on the chest. It was now as firm as concrete, but Misery kept on until he had barked his knuckles and then he appeared to be satisfied. Next he pinched the ghost's muscles as well as he was able—they were as tough as telephone cables now—and then he pulled the ghost's jaw down and began rapping his teeth until they rang like china cups. Misery went and sat down on the bed.

'I don't think I can give you a job,' he said. 'You'd need a place to sleep during the day, and that'd probably mean I'd have to buy another chaff bag and—'

'I never sleep,' said the ghost. 'I'll work day and night if you'll only give me a job. Anything but haunting—'

'I might take you on, but, mind you, I ain't promising,' said Misery, getting back to bed. 'How am I to know you won't get tired of the job and chuck it up after a month and leave me in the lurch?'

'I'll spit me death or take an oath if you like, so long as you give me a job,' said the ghost.

'I'll have to think it over,' said Misery. 'There's a catch in it somewhere. Everybody's out to get you.' He scratched three hairs out of his beard. 'Besides, there's one thing you haven't taken into account. You come here begging me for a job and think that's all there is to it!'

'Oh, what's that?' asked the ghost. 'If I can meet you in any way—'

'Wear and tear,' said Misery. 'Me horses and plough and harvester and shearing gear and pick and shovel and the rest. Stands to reason there'll be wear and tear.'

'I'll be light on things,' said the ghost, screwing up his forehead as he spoke, and starting to waver in a gust that came in the door.

'They all say that,' said Misery. 'Promises are cheap. Besides, I reckon you'll get me some way or t'other.'

'Well, there's one thing,' said the ghost. 'I wasn't going to mention it, but now you've brought it up—'

'I knew it!' said Misery, and he lay down and started pulling up the chaff bag.

'Don't go,' said the ghost. 'It's only a little thing and won't cost you anything. I'd like a ride on your billy-goat. It sounds kind of silly, but when I was alive I wanted to join the Masons, but now I'm a ghost I suppose this is as close as I'll ever get to it.'

'You call that a little thing!' cried Misery loudly, sitting up. 'He's a most valuable goat and you'd be certain to break his back or scare the wits out of him.'

'I'd make myself as light as a feather,' said the ghost. 'And invisible. He'd never know.'

Misery didn't reply for five minutes. 'All right,' he said then, 'as a favour I'll let you work for me for two years, and at the end of it I'll let you ride the goat, though, mind you, it'll have to be a short one.'

'One year,' said the ghost, who was waking up a little to Misery by this time.

'Eighteen months,' said Misery.

'Twelve months,' said the ghost.

'OK, twelve months, though you drive a hard bargain,' said Misery then. 'But, mind you, no women and no beer. You can hook up the team and start the sowing straight away.'

The ghost began working that night, and at the end of three weeks the wheat crop was in. He was a good worker, though Misery growled a bit when he woke up one night and saw the team stopped and the ghost lounging against the trunk of a gum. He got up out of bed and went over to ask what it was about.

'Smoke-oh,' said the ghost. 'I don't smoke but I can't scab too much on the fellows. Otherwise, work's just what I need.'

'I knew there'd be a catch somewhere,' said Misery as he went back to bed.

After the sowing the ghost asked for another job and Misery pointed to a hill covered with granite boulders. 'You could sow another fifty acres if those boulders were grubbed out.'

The ghost went at it and at the end of a month the hill was sown.

'We'll be ruined if there isn't any rain soon,' said Misery, as he put the ghost on to sinking three dams.

'I reckon I might be able to fix the rain, too,' said the ghost. 'I've got some influence.'

'OK,' said Misery, 'but don't get behind with the tank-sinking.'

That night Misery woke up about midnight and saw the ghost wasn't at work. He was still fuming three hours later when the ghost returned, scudding along at about seventy miles an hour, just over the tops of the fences and a few yards ahead of a squall of rain. It rained for a day and a half, and Misery, after saying that the ghost had done it on purpose to get out of the tank-sinking, put the ghost to work mending harness, digging out drains, shoeing horses, repairing machinery, and husking corn.

'I'll be ruined if I don't get some good sun,' said Misery when the ghost went back to the dam-sinking. 'The place is a quagmire. I knew there'd be a catch in it.'

'I might be able to help out there,' said the ghost.

'All right,' said Misery, after a pause, 'but don't overdo it this time and don't leave the job.'

Though the ghost stayed at work the next day, the sun started to get good and hot, the water began to dry up and the crop to grow before their eyes. At the end of a week the ghost had sunk the dams and Misery put him to work to shear the sheep, to break in three sulky horses, and to clear four hundred acres of scrub, saying as he did so, 'I'll be on relief if it doesn't rain soon. I was going to allow a fortnight for these jobs, but I don't want to be a slave-driver—I'll give you a month.'

Misery stayed awake all that night and, though he saw the ghost hard at work and heard the ring of his axe and the whirr of his shears, as he dashed from one job to another, the rain arrived as before, in a squall, and poured for a day and a half.

'No one ever does things right,' Misery told the ghost then. 'The shearing is all held up, the sheep will get foot-rot, and the crop will be ruined if a good baking sun doesn't come along soon. I said there'd be a catch if you came to work for me.'

The ghost fixed the sun the way he had done before. He grafted for Misery for a year and then asked for his wages.

'On the dot!' said Misery then. 'You're all clock-watchers today! I'm a man of my word, however.'

He led the goat out and the ghost looked a bit sheepish, now it had come to the point. He was wearing white mole-skin pants for the occasion.

'I'll claim damages if you harm him in any way,' said Misery. 'Light as a feather and invisible it has to be.'

'OK,' said the ghost, and he frowned and started fluttering in the breeze and then fading. The last thing Misery saw was a misty leg lifting itself over the goat's back.

'Only a hundred yards, mind you—no, fifty,' said Misery.

'As you wish,' said the ghost, but not with a good grace. He was beginning to have enough of Misery by this time.

'And no spurs,' said Misery.

'I can't do anything about them,' said the ghost. 'They're yours and I can't make them vanish.' He was fed up with Misery now, and he dug the spurs in hard. The goat whirled its head to look round. The spurs dug in again. The goat got the wind up when he saw no one on his back and bounded forward. He knocked Misery over and headed for the hut, where he bucked. The ghost was shot forward and he made himself solid to take the fall. He went head first into the hut, knocked it flat, and kept on going. Misery got furious, and then chased the ghost, yelling for him to come back and mend the hut. He panted after the spurs for a couple of miles before they put on a bit of a sprint and left him behind. The goat never came back.

'I was right about there being a catch in it,' said Misery, and put a notice up:

The Tale
of
Caseley Halt

BARRY SUTTON

Jean and I always enjoyed visiting our friends John and Julie Thomas—or, to give respect where it is due, Dr John and Mrs Julie Thomas. Quite apart from his impeccable taste in wine and the blazing warmth of an unstinted fire in his comfortable lounge, John always told a most entertaining line in stories, generally of a supernatural kind. In fact, he told them so well that we often had the utmost difficulty in determining whether or not they were true. The one he recounted to us on this particular night was no exception, and this is how it went.

I suppose (began Dr Thomas) that ghosts are traditionally associated with ruined houses, ancient churchyards and castles in which innocent victims of fate were cruelly done to death. The story I am going to tell involved none of these favoured haunts for the restless departed, yet for all that it was an experience that was to have the greatest significance in my life.

It all happened just over nine years ago at the time Dr Beeching was busily closing most of our branch railway lines. I had to go down to the west country to clear up the affairs of a recently deceased uncle who lived at a place called Throwleigh on the edge of Dartmoor.

So far as I was concerned it was not likely to prove a particularly lucrative excursion—my uncle being a poor man and I not one of his favoured nephews. However, there was nothing for it, so I left instructions with my partner to take the surgery during my absence and to visit a few of my patients, and headed out of London on the A30.

It was late evening when I reached the west country and by that time a drizzle, which had threatened all day, had now developed into a sizeable downpour. I stopped at Exeter for a cup of tea and bought a copy of the *Western Evening Star*, noting with no more than passing interest that they still hadn't found the girl who had been missing on Dartmoor for the past five days.

It was getting dark as I left the city and somehow at Chudleigh I missed my turning and found myself on a by-road signposted: 'Hennock'. It had been my intention to stop the night at Bovey Tracey, but when I eventually came back on to the main road beyond that place I decided I would stop at Moretonhampstead instead.

By this time it was a really foul night and the rain was lashing into the headlights in an almost solid cascade. Suddenly in the blurred yellow light I saw a movement on the road ahead. It resolved itself into the shape of a woman in a light-coloured mackintosh with a red scarf tied over her head. She was waving frantically for me to stop, and I pulled in imagining that she had missed the bus and was wanting a lift. I lowered the near-side window to speak to her and saw that she was much older than I had at first imagined and that strands of grey hair were streaming with rain from under her scarf.

'Thank God it's you, Doctor,' she gasped. 'I'd almost given you up.' She opened the door of the car and sat down abruptly in the passenger seat.

Before I had regained my surprise at the familiarity of her approach, she was giving me directions of some sort which I was apparently expected to follow. We turned off the main road into a narrow lane with dark banks on either side and after a mile crossed over a stone bridge with the swollen torrent of a stream dimly visible below.

'I've made her as comfortable as I can, Doctor, but I think she may have broken her leg. She's complaining of pains in her chest as well.'

As she continued it gradually dawned on me that the woman was referring to her daughter who had, it seemed, met with some kind of accident. I could only assume that she must have caught sight of the BMA badge on the front of the car and sought my help.

After a couple more turnings we arrived in front of a small group of buildings which I recognized as being those of a country railway station. I grabbed my bag from the back of the car and followed the woman through a small white picket gate on to the platform. A single oil lamp hung from the wooden canopy and shed a pale light on to the glazed asphalt and the single pair of glistening rails embedded in the rust-coloured ballast below. A large board announced the name of the place: 'CASELEY HALT'.

We went past the tiny ticket office—now closed—and entered the door of a building at the end of the platform which I guessed was the dwelling house of the station-master. Immediately we were in a small front parlour and, in contrast to the wild night outside, the place was filled with the fug of a coal fire which burnt brightly in the grate.

The room was not much more than ten feet square with cream distempered walls and crowded with bulky furniture. However, the object which attracted my attention was the girl lying on the leather settee. She was about twenty, with fair hair, and there was something about her face which struck me as vaguely familiar. I was sure I had seen that face somewhere before. At the present moment her features were deathly white and her eyes closed.

'It's her foot, Doctor.' The woman had taken off her mackintosh and scarf and had laid them over the back of a chair. She was wearing a brown cardigan with a distinctive cameo brooch at the throat. 'You see, she fell . . .'

I had turned back the rug which covered the girl. Her eyes opened and she gave a brief wry smile of pain as I touched the lacerated tissues. Gently I articulated the foot, noting with satisfaction the absence of the tell-tale crepitus which might have indicated a fractured bone. At last I was satisfied. I turned to the woman.

'Fetch me some warm water and a flannel—oh, and some soap.'

Whilst I was bandaging up the girl's wound I had the first opportunity to talk to her.

'How did it happen?' I asked.

She began to tell me how she had fallen into some soak-away pit behind the station house. I nodded absently. Neither of them looked the kind of people you would normally expect to find at a small out-of-the-way station. I assumed they were the family of some absent member of the railway staff.

There was a sudden imminent rumbling and a train shook the building on its way past outside. I am quite certain about that train. I not only heard it—the clank of the engine and the huff of steam—but I also smelt the oily sulphurous smoke which wafted under the door, and saw the reflected glow of the fire on the window.

The girl was looking better already and some of the colour had returned to her face. I turned to the woman.

'I've got to come this way in the morning so I'll just call in and see how she is. In the mean time I think you should get your own doctor to look at her.' The woman thanked me and, because of the rain, we parted at the door of the house.

The next morning was one of those utter contrasts which we sometimes get in the British Isles. The sky was not only cloudless—it had that scintillating brightness characteristic of a freshly peeled transfer. I had done all I could at Throwleigh and seen my uncle's solicitor at Moretonhampstead. It seemed I would have to go back there again the following week, but for the time being I was enjoying a relaxing ride through the vivid greenness of the Bovey Valley.

I remembered my call of the previous night and found the lane I had to turn down to reach the railway. I crossed the stone bridge. If anything, the river was running higher than when I had glimpsed it the night before. At last I came to the short drive which led up to the station and was surprised to see how weedy and overgrown it was by daylight.

I parked my car and looked around in bewilderment. The station was no longer there. A crumbling ramp was all that remained of the platform. Rotting piles of woodwork marked the little picket fence. With a sense of profound shock and disbelief I picked my way across the litter of dusty bricks where the ticket office had stood and gazed down on to the bleached and empty chippings of the track bed. Weeds—ragwort, coltsfoot and willow herb—sprouted freely between the stones. In both directions small silver birch trees closed over the way that the trains had taken. At my feet I kicked over the rusty grate in which the fire had blazed so merrily the night before. The place was utterly desolate.

At the far end of the platform the one remaining building—a roofless, broken rectangle which could have been an outhouse to the station-master's dwelling—drew me in curiosity. As I drew near I heard a faint moan from beside the building. Crumpled against the wall, with one leg drawn up under her, lay the girl I had attended to the previous evening. Her clothes were sodden and she looked in the final stages of exhaustion.

'Whatever's happened?' was all I could think to ask, but her head just lolled back in a dead faint. It was then I remembered where I'd seen her. Of course—the photograph in the *Western Evening Star*. It was the girl who had been missing on the moors since the previous Tuesday.

I had to come down to the west country the following week and I called in at the hospital to see how she was getting on. I

wanted to make sure she was in good hands, but also I had to resolve a question which had been gnawing at my mind all the week. After a suitable time by her bedside I asked, as casually as I could, 'Do you happen to know a woman—I suppose in her late fifties. The only time I ever saw her she was wearing a light mackintosh and a brown cardigan with a distinctive cameo brooch.'

The girl smiled. 'Yes, certainly I do.' She reached for the small drawer in the bedside cabinet and picked something out. 'Is this the brooch you mean?'

I nodded.

'It belonged to my mother,' the girl continued. 'She died four years ago.'

'Ah yes—of course,' I breathed.

John Thomas folded his hands together and beamed in satisfaction. The firelight glinted on his glasses. Jean and I relaxed in our chairs. Of course, we didn't believe a word of it.

Dr Thomas leaned forward and spoke to his wife. 'Julie, be a dear and fetch that madeira from the sideboard.'

It was then that we noticed. She wore a distinctive cameo brooch, and as she moved across the room we were reminded of her slight limp. They were married nine years ago last month.

Christmas Meeting

ROSEMARY TIMPERLEY

I have never spent Christmas alone before.

It gives me an uncanny feeling, sitting alone in my 'furnished room', with my head full of ghosts, and the room full of voices of the past. It's a drowning feeling—all the Christmases of the past coming back in a mad jumble: the childish Christmas, with a house full of relations, a tree in the window, sixpences in the pudding, and the delicious, crinkly stocking in the dark morning; the adolescent Christmas, with mother and father, the War and the bitter cold, and the letters from abroad; the first really grown-up Christmas, with a lover—the snow and the enchantment, red wine and kisses, and the walk in the dark before midnight, with the grounds so white, and the stars diamond bright in a black sky—so many Christmases through the years.

And, now, the first Christmas alone.

But not quite loneliness. A feeling of companionship with all the other people who are spending Christmas alone—millions of them—past and present. A feeling that, if I close my eyes, there will be no past or future, only an endless present which *is* time, because it is all we ever have.

Yes, however cynical you are, however irreligious, it makes you feel queer to be alone at Christmas time.

So I'm absurdly relieved when the young man walks in. There's nothing romantic about it—I'm a woman of nearly fifty, a spinster schoolma'am with grim, dark hair, and myopic eyes that once were beautiful, and he's a kid of twenty, rather unconventionally dressed with a flowing, wine-coloured tie and black velvet jacket, and brown curls which could do with a taste of the barber's scissors. The effeminacy of his dress is belied by his features—narrow, piercing, blue eyes, and arrogant, jutting nose and chin. Not that he looks strong. The skin is fine-drawn over the prominent features, and he is very white.

He bursts in without knocking, then pauses, says: 'I'm so sorry. I thought this was my room.' He begins to go out, then hesitates and says: 'Are you alone?'

'Yes.'

'It's—queer, being alone at Christmas, isn't it? May I stay and talk?'

'I'd be glad if you would.'

He comes right in, and sits down by the fire.

'I hope you don't think I came in here on purpose. I really did think it was my room,' he explains.

'I'm glad you made the mistake. But you're a very young person to be alone at Christmas time.'

'I wouldn't go back to the country to my family. It would hold up my work. I'm a writer.'

'I see.' I can't help smiling a little. That explains his rather unusual dress. And he takes himself so seriously, this young man! 'Of course, you mustn't waste a precious moment of writing,' I say with a twinkle.

'No, not a moment! That's what my family won't see. They don't appreciate urgency.'

'Families are never appreciative of the artistic nature.'

'No, they aren't,' he agrees seriously.

'What are you writing?'

'Poetry and a diary combined. It's called *My Poems and I*, by Francis Randel. That's my name. My family says there's no point in my writing, that I'm too young. But I don't feel young. Sometimes I feel like an old man, with too much to do before he dies.'

'Revolving faster and faster on the wheel of creativeness.'

'Yes! Yes, exactly! You understand! You must read my work some time. Please read my work! Read my work!'

A note of desperation in his voice, a look of fear in his eyes, makes me say:

'We're both getting much too solemn for Christmas day. I'm going to make you some coffee. And I have a plum cake.'

I move about, clattering cups, spooning coffee into my percolator. But I must have offended him, for, when I look round, I find he has left me. I am absurdly disappointed.

I finish making coffee, however, then turn to the bookshelf in the room. It is piled high with volumes, for which the landlady has apologized profusely: 'Hope you don't mind the books, Miss, but my husband won't part with them, and there's nowhere else to put them. We charge a bit less for the room for that reason.'

'I don't mind,' I said. 'Books are good friends.'

But these aren't very friendly-looking books. I take one at random. Or does some strange fate guide my hand?

Sipping my coffee, I begin to read the battered little book, published, I see, in spring, 1852. It's mainly poetry—immature stuff, but vivid. Then there's a kind of diary. More realistic, less affected. Out of curiosity, to see if there are any amusing comparisons, I turn to the entry for Christmas day, 1851. I read:

'My first Christmas day alone. I had rather an odd experience. When I went back to my lodgings after a walk, there was a middle-aged woman in my room. I thought, at first, I'd walked into the wrong room, but this was not so, and later, after a pleasant talk, she—disappeared. I suppose she was a ghost. But I wasn't frightened. I liked her. But I do not feel well tonight. Not at all well. I have never felt ill at Christmas before.'

A publisher's note followed the last entry:

Francis Randel died from a sudden heart attack on the night of Christmas day, 1851. The woman mentioned in this final entry in his diary was the last person to see him alive. In spite of requests for her to come forward, she never did so. Her identity remains a mystery.

An Apple
for
Miss Stevenson

MICHAEL VESTEY

They knew January was a bad month to go house-hunting but they had no alternative. They had already sold their own house. Kent looked more like the backyard than the garden of England. Still, the Sheldons reasoned, it was the same bleak landscape in all the Home Counties and it wouldn't put them off living in the country.

Phillip Sheldon and his wife Elizabeth were both Londoners but they had made their decision to move out, like so many young couples before them. They now felt strangers in the city, and they disliked the noise, the discomfort, the poor schooling, and the growing shabbiness. Besides, there were the children to think of, or, rather, their one child, a little girl of five called Emma. Far better for her to grow up in the country than the town.

That is how they came to be standing at the lichen-cloaked gate to Rose Cottage on the edge of a village in the heart of the Kentish hop fields. Phillip shivered in the coldness of the New Year. The sky was the colour of the wood-smoke that spurted from chimney pots in the village. The trees in the rambling garden were bent like arthritics in the sharp wind. Gamekeeper's weather, thought Phillip. He wondered if he would miss their warm, centrally heated, purpose-built townhouse in the suburbs.

Elizabeth looked up at her husband and smiled. 'Well, it's big,' she said.

Phillip frowned. 'Yes . . . are you sure we've got the right place?' There was no For Sale sign evident. 'It seems too big for a cottage.'

Without knowing what the inside was like, Phillip felt instinctively that the house was a bargain. He looked down at the estate agent's particulars on a sheet of white paper flapping in his cold hands.

Elizabeth said: 'I'm amazed it hasn't gone before now. It must be awful inside.'

'Right, let's go in and see the dreadful truth,' he said, opening the gate and walking up the drive. Elizabeth was about to follow him when she felt her arm jerked backwards. It was Emma. The little girl stood firmly the other side of the gate. Her pale face could just be seen peeping through the hood of her black duffle coat and the thickly entwined scarf.

'No,' cried Emma. 'No.'

Phillip was half-way up the drive, unaware of his daughter's reluctance to follow him, when he suddenly realized he was alone. He turned, a tall, thin, pin-striped figure, wisps of dark hair blowing across his spectacles. 'Well, come on, then,' he yelled, stamping his feet. 'It's chilly out here.'

'It's Emma,' his wife replied. 'She won't come.'

'Oh God! What a time to throw a tantrum,' muttered Phillip. She'd been so good on the journey down and even at the previous house they had viewed. He could see that Emma clearly did not want to budge. What he could see of her tiny white face was set hard with determination, an expression both he and Elizabeth had learned to identify. The last time she had behaved in this way was on her first day at school.

Elizabeth crouched and put her face close to Emma's. 'Darling, we won't be long, this is the last house we'll see today. And then we can all go and have tea somewhere. Won't that be nice?'

Emma's chin trembled and her face began to crumple. She started to cry.

Phillip became impatient. 'Can't she play in the garden while we're inside?' he shouted.

'No,' replied Elizabeth. 'If she's going to live here, then she must see it first.'

'Oh well, I'm going on. You join me later.' Phillip strode up towards the house which sat on a small hill at the highest point of the garden. Weeds sprouted through the worn gravel and scattered shingle of the neglected drive. 'The garden was once a feature of the house,' he said to himself, repeating the words in the brochure. 'Not any more it isn't.' He glanced at the thick bushes, matted together like uncombed animal fur, the deep-pile lawn which would take several mowings to recover its health, and the brambles snaking across what had once clearly been a vegetable patch. The summer would be an energetic time!

A few yards from the house he stood with hands on hips and looked it over. It was timber-framed and clad in white-painted weatherboarding, some of which was peeling. The window and door frames were painted black to provide a sharp contrast. Unpruned climbing plants covered the downstairs windows, making it impossible to see inside. There were solid brick chimneys at both ends of the house, between which the sand-coloured tiled roof gently undulated.

He turned the key in the lock and with some difficulty pushed open the front door. Inside it felt damp and almost colder than it was outside. He could smell the decay caused by the moisture in the floorboards and plaster. It came as no surprise to him. After all, the agent had made it clear the house had been empty for two years.

After passing through a small lobby he found himself in what appeared to be the main sitting-room. It had wide, oak floorboards in reasonably good condition, and a huge inglenook fireplace, high enough for a man his height, six feet, to stand up in without bumping his head.

'Isn't it lovely?' said a voice behind him. Elizabeth stood in the doorway holding Emma by the arm. Neither parent noticed that Emma's face had turned a sickly grey.

'I don't know what's the matter with her. I had to promise all sorts of bribes to get her up here. There, Emma!' she said, tugging at the little girl's arm. 'What do you think?'

Emma stared with deep curiosity in the direction of the inglenook fireplace. She said nothing. Her parents shrugged, and moved towards the kitchen, which turned out to be a high-ceilinged room with black-painted oak beams running from end to end. One side of the ceiling sloped downwards to follow the line of the cat's slide roof at the rear of the house.

Upstairs there were five bedrooms, including a small room in the roof. Elizabeth was attracted by the character of the house with its oak floors and beams; Phillip by the size of the rooms and the feeling of space in a house three hundred years old. An additional bonus was that he had bumped his head only once—on a low door frame.

'Yes,' Elizabeth murmured. 'We must have this.'

'Subject to survey and contract,' said Phillip, who, as a solicitor, was not going to allow sentiment to overwhelm him entirely. They both laughed. 'Come on, then,' said Phillip. 'Let's go back and make an offer.'

When they reached the front door they realized that Emma was not with them any more. They called out and listened. Not a sound, apart from the wind. Phillip became aware of how quiet it would be living there and he wondered, fleetingly, if he would be able to cope after the city.

'I'll go and find her,' said Elizabeth, moving towards the room with the inglenook. There she spotted Emma seated in the gloom of the fireplace engrossed in something, she couldn't see what.

'Oh, there you are, darling. Come on, we're off.'

Emma seemed startled. She glanced up at her mother, and in the poor light Elizabeth noticed a familiar expression of guilt on Emma's face. However, Elizabeth was anxious to return to the estate agent's. She held out a hand. Emma nodded, smiled and then skipped towards the front door. Surprised but also pleased at the sudden change of attitude in her daughter, Elizabeth said: 'That's better.'

At the estate agent's, the senior partner Mr Wilcox showed little enthusiasm when the Sheldons told him they wanted to buy Rose Cottage. 'You *liked* it then?' he asked cautiously, seated behind his leather-topped desk.

Phillip nodded. 'It's got more than enough for a growing family, and bags of character, of course.'

'Yes, it's certainly roomy,' replied Mr Wilcox, unscrewing the top of his fountain pen. 'It was once a school, you know.'

'A school!' both exclaimed in unison. 'It's not that big, surely?' questioned Phillip.

'No, but it was big enough to be a small prep school. It was called St Anne's.' He smiled. 'Occasionally, the local vet and the supermarket manager stop and stare over the garden gate with nostalgic looks on their faces.'

Elizabeth and Phillip gave Mr Wilcox their full attention as he gave them a brief history of the house. It was a farmhouse until the turn of the century when it was bought by a wealthy widow who lived there until the outbreak of the Second World War. She sold it to two middle-aged sisters, Amy and Charlotte Stevenson, both spinsters who ran a prep school in Dover.

Rather than risk the bombs, the sisters searched inland for suitable premises to house their school. That's how they found Rose Cottage, explained Mr Wilcox, though it wasn't called that then. Because of its size the school could cater only for about twenty-five to thirty pupils with a few boarders. In their spare time the sisters, both amateur horticulturists, transformed the garden from an ordinary patch of lawn and shrubbery into what Mr Wilcox fondly recalled as a 'scented paradise' of rare flowers and plants. 'It was lovely then,' he sighed.

'Pretty red roses,' said Emma suddenly.

Mr Wilcox glanced across at Emma who was sitting on a chair by the window. He seemed surprised. 'Yes, that's right, young lady. Miss Amy Stevenson's roses were famous in this area, especially the red ones.'

'I didn't see any roses,' said Phillip. Elizabeth shook her head.

Mr Wilcox laughed. 'All roses are red to children,' he said with a chuckle. But when he had spoken the cheerfulness gave way to a frown.

'By the way,' said Elizabeth. 'What happened to the old ladies?'

'Ah, yes.' Mr Wilcox clasped his hands together on the desktop, pursed his lips and squinted slightly through his spectacles. 'Well, they both passed on. Rather sad really: St Anne's became too much for them. In the last years they could manage only elocution lessons. That's when the school ceased and they called it Rose Cottage instead. First Miss Charlotte went, in hospital, and then Miss Amy, about a year later, I think. The house has been empty since then, as you could see from its neglect.'

'I'm surprised you haven't been able to sell it before now; it's not in that bad condition,' said Phillip.

Mr Wilcox agreed. After an initial hesitation he gave a short nervous laugh. 'Well, the fact is . . . I mean, I don't personally believe in these things myself but . . . well, some say the house is, well, haunted.'

They both laughed. 'Haunted! I've always wanted to live in a haunted house,' cried Elizabeth, clapping her hands together with amusement.

'You didn't notice anything, then?' Mr Wilcox asked almost timidly.

Phillip answered: 'Such as?'

'Well, I don't know. Some people who have been round it say the house has a definite atmosphere. In fact, to be quite frank with you, I showed a young couple like yourselves round it recently and afterwards they told me, and I quote, that it gave them the creeps. Couldn't see it myself but there we are. Before you decide, you ought to talk to a neighbour, a retired doctor called Hadley, he's—'

'Well, I don't think I'll bother with that, thank you,' Phillip interrupted sharply. 'Like you, I don't believe all that stuff.' He looked at his watch and said it was time they returned to London. He made an offer for Rose Cottage which was considerably less than the asking price. Even Phillip was surprised when Mr Wilcox replied instantly and without thought: 'Accepted.'

'Don't you want to consult the vendor?' asked Phillip.

'No, it's an executors' sale. They've left it to me. Glad to get it off the books, if you really want to know.'

As they drove back to London they said little, working out in their minds their respective plans for Rose Cottage. As they approached the yellow street lighting of the outer suburbs, Emma spoke for the first time on the journey.

'Some of the little children brought red roses for the school and Miss Stevenson put them in pots and they grew again in the garden. That's why she had such lovely roses.'

'Oh really,' answered Elizabeth, preoccupied with wallpapers for the upstairs bedrooms. 'Is that what Mr Wilcox told you? What a nice story.' Her thoughts returned to the problem of the main bedroom. It might have to be painted white, paper might not go with oak beams . . .

As they slowed for traffic lights Phillip wondered what sort of mortgage he would succeed in raising on such an old house.

The Sheldons were an efficient couple. As a solicitor he was able to carry out his own conveyancing on the house. In the meantime, she was responsible for organizing builders. She made several trips to the house to supervise the work which was duly completed a mere two weeks after the target date. After staying with friends for a few weeks, they moved into Rose Cottage.

This time, Emma displayed no reluctance to return to the house; if anything, she seemed to look forward to it. She also started at the village school.

About a fortnight later, on a cold morning in spring, over breakfast in the warm kitchen, Emma said: 'Don't want to go to school today.'

Phillip continued reading his newspaper, while Elizabeth sewed a name-tag into Emma's coat. Emma sat at the table pouting. Her rust-coloured hair that fell to her shoulders gleamed in the weak sunlight from the kitchen window. She repeated herself and finally elicited a response.

'Why not?' asked her mother, continuing to sew.

'Miss Stevenson told me not to,' came the brief reply.

'Who is Miss Stevenson?' asked Elizabeth, biting the thread with her teeth and wrenching away the loose end. 'There! That's another one done.' She held up the coat and admired her handiwork.

'Miss Stevenson lives *here*,' replied Emma impatiently. 'You know.'

'What on earth are you talking about, child?' said Elizabeth, taking an interest for the first time. 'A Miss Stevenson used to live here but she's, well, she went to heaven, didn't she?'

'No she didn't,' Emma shouted. Her face began to match the colour of her hair. She was normally quite pale. 'She didn't, she didn't, she's here!'

At last, Phillip lowered his newspaper. 'What's the matter with her now?'

Elizabeth put her arm around Emma's shoulders. 'Now, darling, you are going to school today, and that's that. I don't want to hear any more nonsense about Miss Stevenson. Has someone at school been telling you stories about her?'

A look of panic crossed Emma's face. She gulped, as if for air, jumped off her chair and ran towards the stairs.

'How extraordinary,' said Phillip, looking at his wife. 'Has she ever done that before?'

'No, not since her first day at that other school. What are we going to do?'

'Send her to school, of course. There's nothing wrong with her.'

Elizabeth thought for a moment. 'The trouble is, Phillip, she does seem to have a thing about this Miss Stevenson. I've heard her holding conversations with someone in her room, and sometimes by the fireplace in the sitting-room.'

'Oh, all children do that,' said Phillip scornfully.

'Yes, that's what I thought. But something's odd. She recited her two-times table the other day in her room. She didn't know I was listening the other side of the door. And she knows her alphabet, and I know for certain that she hasn't got round to these things yet at school.'

Phillip wasn't convinced. 'It's play-acting,' he said, standing up. 'She just doesn't want to go to school. We'd better put a stop to it now.' He started to walk upstairs, beckoning Elizabeth to follow him.

The white-painted door to Emma's room was closed. Phillip was about to turn the handle when he heard a child's voice. He listened carefully.

Emma was saying: 'They told me I *had* to go to school.'

There was a long pause. Phillip and Elizabeth held their heads against the door as closely as possible without making a noise.

Emma began speaking again: 'No, I won't, I promise.'

Her parents heard a chair scraping across the floorboards, and then Emma spoke again: 'Once two is two, two twos are four, three twos are six, four twos are eight . . .'

Phillip turned the handle and pushed his way into the room. Emma was sitting on the bed surrounded by exercise books. She held a pen in her hand. No one else was in the room. Emma smiled innocently. 'Hallo, Daddy, I'm just doing my tables.'

'Yes, so I can see. Now put your coat on and Mummy will take you to school.'

'But I'm doing my lessons,' replied Emma with a frightened look on her face.

Phillip advanced towards the bed, reached down and smacked Emma on the leg. 'You'll do as you're told.'

Emma began crying. 'Miss Stevenson said I wasn't to.' She put a hand up to her face as if to staunch the flow of tears.

Phillip knelt by the bed and said firmly: 'There is no Miss Stevenson any more, do you understand, Emma?'

Emma pointed to a chair by the window. 'That's Miss Stevenson,' she said, quietly sobbing. 'She teaches me things.'

Phillip got up and walked over to the chair. He swept an arm across the seat. 'There, you see, nothing. There is no one there. It's just pretend. Now do you believe us?'

Emma's eyes widened. 'She is there, she is,' she repeated.

Phillip was becoming exasperated. He looked at his watch. 'Damn! I've missed my train.' He wagged a finger at Emma. 'You are going to school if I have to drag you there.'

Emma stared fearfully at the chair by the window. She seemed for a moment to be filled with terror. She began gasping for breath, her tiny chest heaving as she fought for air. The whites of her normally grey-blue eyes bulged slightly and the colour drained from her face. To her parents' astonishment she began to convulse.

'Christ, she's having a fit,' yelled Phillip. He urged Elizabeth to run down the lane to fetch their neighbour, Dr Hadley. Meanwhile, Phillip tried to loosen Emma's jersey. She had successively gone white, yellow and purple. Now she was grey.

Suddenly, he was pushed aside by a scruffily dressed man who stood over Emma and felt her pulse. She was still shaking but not so violently. Dr Hadley pulled a blanket over Emma, turned and said: 'I'll be downstairs to see you in a moment.'

Phillip and Elizabeth took the hint and went downstairs to the sitting-room. After what seemed an inordinately long time—though it was only ten minutes—Dr Hadley joined them. He was wearing a worn tweed jacket, rough trousers and mud-stained boots. He was almost bald but had a pleasant weather-beaten face and a friendly manner.

'You caught me gardening,' he said, as if reading their thoughts. 'She's all right now. She's asleep. Has she had fits for long?'

'Never,' replied Elizabeth.

'Any history of epilepsy?'

'Good heavens no!' exclaimed Phillip. 'Nothing like that.'

'We had a little chat before she went off to sleep,' said Dr Hadley.

'Did she mention how it all started?' asked Phillip.

Dr Hadley nodded. 'Very interesting, very interesting.'

'You don't believe all this nonsense, do you? Surely it's absurd to think this house is haunted by Miss Stevenson?' Phillip now seemed angry.

Dr Hadley smiled. 'Maybe, but I tend to keep an open mind.'

Phillip failed to notice this subtle rebuke and might have been rude had not Dr Hadley continued: 'You see, I knew Miss Stevenson well. She was a very strong-willed woman. If the house is haunted I can't say I'm surprised that it's her and not her sister. There are several people in the village who say they've felt her presence though no one has ever seen her—until now.'

Phillip was startled. 'So you do believe her?'

'I'm not saying that, but your daughter tells me she is here and that she gives her lessons. And there are one or two things she told me about the sisters that could only have come from them.'

'Such as, Doctor?' Elizabeth wanted to know.

'Well, Emma told me that there had once been a little girl at the school just like her, with red hair and even the same colour eyes. Indeed there was. I remember the girl quite well, I once gave her a jab for something or other. There's a quite astonishing likeness. Now, only Miss Stevenson could have told her that. She also mentioned a little boy who kept climbing the hop poles in the next field and how the farmer chased him away. Now that happened twenty years ago. Did you know any of that?'

Phillip was shocked for a moment. He glanced at Elizabeth who said: 'There was that story about the red roses, Phillip. Do you remember, in the car on our way back to London after we'd bought the house? I didn't think anything of it at the time but thinking back, how on earth could she have known about the roses?'

'The roses the children brought and which Miss Stevenson quickly re-potted and later planted in the garden?' asked Dr Hadley.

'Yes,' Elizabeth answered.

'They were famous for miles around, those roses. They came from the children of a local rose grower. Miss Stevenson couldn't afford to buy his roses herself, so she used the cuttings the children brought. She was no fool.'

Phillip now felt less certain about it all. He had never believed in ghosts, and still didn't, but he was no longer quite so confident in his scepticism. 'If all this is true,' he said slowly, 'then Emma must have first met Miss Stevenson the day we saw over the house.'

'It does explain a lot of things we were too busy to notice at the time,' said Elizabeth.

After a pause, Dr Hadley asked: 'What are you going to do? These fits are no good for her at all.'

'Well, it seems to me,' replied Phillip thoughtfully, 'that we have three alternatives: we can move house; we can take her to a psychiatrist, or,' he paused for a moment, 'or, we can bring in an exorcist.'

Rather than uproot themselves again they decided to take Emma to see a psychiatrist in London. She spent three sessions with him, all of which failed to remove the notion of Miss Stevenson inhabiting Rose Cottage as a ghost. The psychiatrist said helplessly that he had never before met a child with such fixed ideas, and that perhaps Rose Cottage was haunted after all. He recommended an exorcist.

This led the Sheldons to their local vicar, a man brimming with sympathy, but offering little hope. 'I knew Miss Stevenson well,' he told them. 'A dear old soul. I can't believe she haunts the place, though I have heard tales, gossip, you know.'

When they persisted he agreed to ask a clergyman in another parish who specialized in exorcism. 'A clear case of possession by an evil spirit must be proved, though,' he warned. 'He'll also have to get the permission of the bishop.'

After the late, cold spring, summer eventually arrived. Phillip worked in the garden at weekends and evenings clearing the undergrowth, digging the vegetable patch and uncovering the famous roses. Phillip had to admit that in full bloom there were some magnificent specimens. Emma helped him sometimes as best she could. She had not been to school for two months. Illness had been the excuse. Now that it was summer she spent many days in a wild part of the garden that she called her own. She had demanded that this area should be left untouched. The wild meadow grass grew high there and Emma could play in her secret garden without being overlooked.

Although Emma had not been to school she was still receiving an education. She allowed her mother to teach her at home but only those subjects not covered by Miss Stevenson.

Reluctantly her parents went along with this, fearful of another scene like that in the spring. But they found they were no longer sleeping as soundly as they did once. Elizabeth needed sleeping tablets for the first time in her life, and Phillip's work began to suffer.

On warm days Emma took her lessons with Miss Stevenson hidden from view in her own patch of the garden. This she would do punctually at nine with a break for lunch at twelve-thirty, returning to lessons at two until four o'clock. Sometimes Emma would take Miss Stevenson a small gift like an apple.

One morning, as Phillip was about to leave for the station, a letter arrived from the local vicar. He regretted that the bishop had refused permission for an exorcism as it was not felt that the little girl or the house were in any way possessed. Instead, he advised medical treatment.

Phillip slumped into an armchair. 'Well, that's it then,' he said dejectedly. 'We'll have to move again.'

Elizabeth agreed, and without Emma realizing, arrangements were made to put the house on the market. Two days later, Mr

Wilcox arrived unexpectedly to view the house. They were surprised to see him as the intention had been to hide the visit from Emma. Mr Wilcox strode into the sitting-room and gazed around him admiringly.

'You have done a lot to the old place,' he said with a broad smile. 'Country life too dull for you, eh?'

Before they could answer, Emma appeared at the doorway from the kitchen. Mr Wilcox grinned at her. 'Hallo, little girl, I can't remember your name but I remember you very well. With that lovely red hair of yours.'

'Emma, say hallo to Mr Wilcox,' said Elizabeth wearily, at the same time wondering how her daughter would react.

Emma said nothing but glared at Mr Wilcox who seemed taken aback by such obvious hostility. Phillip decided to tell the truth. 'Emma, you remember Mr Wilcox don't you? He sold us Rose Cottage. Well, now we've decided to move again, probably back to London. It will be fun to see your old friends again won't it?'

But there was no conviction in his face and Emma could see through it. After staring at nothing in particular in front of her for a few moments she turned on her heels and rushed from the room. They could hear her clumping upstairs.

Mr Wilcox tried to reassure. 'Moving always unsettles them at that age,' he said hastily. 'Now let's get down to—'

He was interrupted by a piercing scream from above, a cry that seemed far too loud and penetrating for a child of five.

'Emma!' shouted Elizabeth, and ran towards the stairs, followed by Phillip and Mr Wilcox. They found Emma screaming hysterically on the bed, turning and twisting her frail little body as if trying to wriggle free from some awful bondage.

'Oh God!' said Phillip, running a hand through his hair. 'Get Dr Hadley.'

Mr Wilcox disappeared, and returned a few minutes later with the retired doctor who gripped Emma's shoulders and shouted: 'It's all right, you're not leaving here. Understand? You're not going away. You can stay with Miss Stevenson.'

Emma seemed, through her hysterics, to comprehend. Slowly, her body ceased its feverish activity; apart from an occasional twitch, she seemed to calm. Her eyes opened and to her parents' relief she managed a weak smile. She began sucking her thumb and turned over on her side. Within seconds she was asleep.

All four looked at each other. Mr Wilcox stood rooted to the spot, his face dripping with perspiration. They left Emma sleeping and crept downstairs. In the sitting-room Dr Hadley apologized. 'I'm afraid it was the only quick way I could think of getting her to stop. I must confess I was surprised that it worked so rapidly.'

'But does that mean we can't move now?' asked Phillip, with a note of desperation in his voice.

'I'd leave it for a bit if I were you,' advised Dr Hadley. 'She might grow out of it, you never know.'

A day later, after much discussion, the Sheldons abandoned the idea of leaving Rose Cottage. They resigned themselves to their predicament. They would have to compromise. After breakfast, Elizabeth stood up, looked at her watch and said: 'Come on, Emma, you're late for your lessons. Miss Stevenson will be cross.'

Dual Control

ELIZABETH WALTER

You ought to have stopped.'

'For God's sake, shut up, Freda.'

'Well, you should have. You ought to have made sure she was all right.'

'Of course she's all right.'

'How do you know? You didn't stop to find out, did you?'

'Do you want me to go back? We're late enough as it is, thanks to your fooling about getting ready, but I don't suppose the Bradys'll notice if we're late. I don't suppose they'll notice if we never turn up, though after the way you angled for that invitation...'

'That's right, blame it all on me. We could have left half an hour ago if you hadn't been late home from the office.'

'How often do I have to tell you that business isn't a matter of nine to five?'

'No, it's a matter of the Bradys, isn't it? You were keen enough we should get asked. Where were you anyway? Drinking with the boys? Or smooching with some floozie?'

'Please yourself. Either could be correct.'

'If you weren't driving, I'd hit you.'

'Try something unconventional for a change.'

'Why don't you try remembering I'm your wife—'

'Give me a chance to forget it!'

'—and that we're going to a party where you'll be expected to behave.'

'I'll behave all right.'

'To me as well as to other women.'

'You mean you'll let me off the leash?'

'Oh, you don't give a damn about *my* feelings!'

'Look, if it hadn't been for you, I should have stopped tonight.'

'Yes, you'd have given a pretty girl a lift if you'd been on your own. I believe you. The trouble is, she thought you were going to stop.'

'So I was. Then I saw she was very pretty, and—Christ, Freda, you know what you're like. I've only got to be polite to a woman who's younger and prettier than you are—and believe me, there are plenty of them—and you stage one of your scenes.'

'I certainly try to head off the worst of the scandals. Really, Eric, do you think people don't know?'

'If they do, do you think they don't understand why I do it? They've only got to look at you ... That's right, cry and ruin that fancy make-up. All this because I *didn't* give a pretty girl a lift.'

'But she signalled. You slowed down. She thought you were going to ...'

'She won't jump to conclusions next time.'

'She may not jump at all. Eric, I think we ought to forget the Bradys. I think we ought to go back.'

'To find Cinderella has been given a lift by Prince Charming and been spirited away to the ball?'

'She was obviously going to a party. Suppose it's to the Bradys' and she's there?'

'Don't worry, she couldn't have seen what we looked like.'

'Could she remember the car?'

'No. She didn't have time.'

'You mean she didn't have time before you hit her.'

'God damn it, Freda, what do you expect me to do when a girl steps in front of the car just as I decide—for your sake—I'm not stopping? It wasn't much more than a shove.'

'It knocked her over.'

'She was off balance. It wouldn't have taken more than a touch.'

'But she fell. I saw her go backwards. And I'm sure there was blood on her head.'

'On a dark road the light's deceptive. You saw a shadow.'

'I wish to God I thought it was.'

'Look here, Freda, pull yourself together. I'm sorry about it, of course, but it would make everything worse to go back and apologize.'

'Then what are you stopping for?'

'So that you can put your face to rights and I can make sure the car isn't damaged.'

'If it is, I suppose you'll go back.'

'You underestimate me, as usual. No, if it is I shall drive gently into that tree. It will give us an excuse for arriving late at the Bradys' and explain the damage away.'

'But the girl may be lying there injured.'

'The road isn't that lonely, you know, and her car had obviously broken down. There'll be plenty of people willing to help a damsel in distress . . . Yes, it's as I thought. The car isn't even scratched. I thought we might have a dent in the wing, but it seems luck is on our side. So now, Freda old girl, I'll have a nip from that flask you've got in your handbag.'

'I don't know what you mean.'

'Oh yes you do. You're never without it, and it needs a refill pretty often by now.'

'I can't think what's come over you, Eric.'

'Call it delayed shock. Are you going to give it me or do I have to help myself?'

'I can't imagine—Eric, let go! You're hurting!'

'The truth does hurt at times. Do you think I didn't know you had what's called a drinking problem? You needn't pretend with me.'

'It's my money. I can spend it how I choose.'

'Of course, my love. Don't stop reminding me that I'm your pensioner, but thanks anyway for your booze.'

'I didn't mean that. Oh Eric, I get so lonely, you don't know. And even when you're home you don't take any notice of me. I can't bear it. I love you so.'

'Surely you can't have reached the maudlin stage already? What are the Bradys going to think?'

'I don't give a damn about the Bradys. I keep thinking about that girl.'

'Well, I give a damn about the Bradys. They could be important to me. And I'm not going to ruin a good contact because my wife develops sudden scruples.'

'Won't it ruin it if they know you left a girl for dead by the roadside?'

'Maybe, but they won't know.'

'They will. If you don't go back, I'll tell them.'

'That sounds very much like blackmail, and that's a game that two can play.'

'What do you mean?'

'Who was driving the car, Freda?'

'You were.'

'Can you prove that?'

'As much as you can prove that I was.'

'Ah, but it's not as simple as that. Such an accusation would oblige me to tell the police about your drinking. A lot of unpleasant things would come out. I should think manslaughter is the least you'd get away with, and that could get you five years. Because please note that apart from that swig I am stone cold sober, whereas your blood alcohol is perpetually high. In addition, you're in a state of hysteria. Who d'you think would be believed—you or I?'

'You wouldn't do that, Eric. Not to your wife. Not to me.'

'Sooner than I would to anyone, but it won't come to that, will it, my dear?'

'I've a good mind to—'

'Quite, but I should forget it.'

'Eric, don't you love me at all?'

'For God's sake, Freda, not that now, of all times. I married you, didn't I? Ten years ago you were a good-looking thirty—'

'And you were a smart young salesman on the make.'

'So?'

'You needed capital to start your own business.'

'You offered to lend it to me. And I've paid you interest.'

'And borrowed more capital.'

'It's a matter of safeguarding what we've got.'

'What we've got. That's rich! You hadn't a penny. Eric, don't start the car like that. You may not be drunk but anyone would think you are, the way you're driving. No wonder you hit that girl. And it wasn't just a shove. I think you've killed her.'

'For God's sake, Freda, shut up!'

'Well, it was a good party, wasn't it?'

'Yes.'

'Moira Brady's a marvellous hostess.'

'Yes.'

'Jack Brady's a lucky man. We ought to ask them back some time, don't you think?'

'Yes.'

'What's got into you? Cat got your tongue? You're a fine companion. We go to a terrific party and all you can say is Yes.'

'I'm thinking about that girl.'

'She was all right, wasn't she? Except for some mud on her dress. Did she say anything about it?'

'She said she'd fallen over.'

'She was speaking the literal truth. Now I hope you're satisfied I didn't hurt her.'

'She certainly looked all right.'

'You can say that again. Life and soul of the party, and obviously popular.'

'You spent enough time with her.'

'Here we go again. Do you have to spend the whole evening watching me?'

'I didn't, but every time I looked, you were with her.'

'She seemed to enjoy my company. Some women do, you know.'

'Don't torment me, Eric. I've got a headache.'

'So have I, as a matter of fact. Shall I open a window?'

'If it isn't too draughty . . . What was the girl's name?'

'Gisela.'

'It suits her, doesn't it? How did she get to the Bradys'?'

'I didn't ask.'

'It's funny, but I never saw her go.'

'I did. She left early because she said something about her car having engine trouble. I suppose someone was giving her a lift.'

'I wonder if her car's still there?'

'It won't be. She'll have got some garage to tow it away.'

'Don't be too sure. They're not so keen on coming out at nights in the country, unless something's blocking the road.'

'I believe you're right. That's it, isn't it—drawn up on the grass verge.'

'Yes. And Eric, that's her. She's hailing us.'

'And this time I'm really going to stop.'

'What on earth can have happened?'

'It looks like another accident. That's fresh mud on her dress.'

'And fresh blood on her head! Eric, her face is all bloody!'

'It can't be as bad as it looks. She's not unconscious. A little blood can go a very long way. Just keep calm, Freda, and maybe that flask of yours will come in handy. I'll get out and see what's up . . . It's all right, Gisela. You'll be all right. It's me, Eric Andrews. We met at the Bradys' just now. My dear girl, you're in a state. What in God's name happened? Has someone tried to murder you? Here, lean on me . . .'

'Eric, what's the matter? Why have you left her alone? Gisela . . .'

'Christ, Freda, shut that window! And make sure your door's locked.'

'What is it? You look as if you'd seen a ghost.'

'She *is* a ghost . . . Give me that flask . . . That's better.'

'What do you mean—a ghost?'

'There's nothing there when you go up to her. Only a coldness in the air.'

'But that's nonsense. You can't see through her. Look, she's still standing there. She's flesh and blood—blood certainly.'

'Is there any blood on my hand?'

'No, but it's shaking.'

'You bet it is. So am I. I tell you, Freda, I put out my hand to touch her—I *did* touch her—at least, I touched where she was standing—but she's got no body to touch.'

'She had a body at the Bradys'.'

'I wonder.'

'Well, you should know. You hung around her all the evening, making a spectacle of yourself.'

'I never touched her.'

'I'll bet it wasn't for want of trying.'

'Now I think of it, nobody touched her. She always seemed to stand a little apart.'

'But she ate and drank.'

'She didn't eat. She said she wasn't hungry. I don't remember seeing a glass in her hand.'

'Rubbish, Eric. I don't believe you. For some reason you don't want to help her. Are you afraid she'll recognize the car?'

'She has recognized it. That's why she's there. We—we must have killed her on the way to the party that time when we nearly stopped.'

'You mean when *you* nearly stopped. When you hit her. Oh God, what are we going to do?'

'Drive on, I think. She can't hurt us.'

'But she could get inside the car.'

'Not if we keep the doors locked.'

'Do you think locked doors can keep her out? Oh God, I wish I'd never come with you. Oh God, get me out of this. I never did anything. I wasn't driving. Oh God, I'm not responsible for what he does.'

'Oh no, you're not responsible for anything, are you, Freda? Does it occur to you that if it hadn't been for your damned jealousy I should have stopped?'

'You've given me cause enough for jealousy since we were married.'

'A man's got to get it somewhere, hasn't he? And you were pretty useless—admit it. You couldn't even produce a child.'

'You're heartless—heartless.'

'And you're spineless. A sponge, that's all you are.'

'I need a drink to keep me going, living with a bastard like you.'

'So we have to wait while you tank up and make ourselves late for the Bradys'. Do you realize, if we'd been earlier we shouldn't have seen that girl?'

'It's my fault again, is it?'

'Every bloody thing's your fault. I could have built up the business a whole lot faster if you'd put yourself out to entertain a bit. If I'd had a wife like Moira Brady, things would be very different from what they are.'

'You mean you'd make money instead of losing it.'

'What do you mean—losing it?'

'I can read a balance sheet, you know. Well, you're not getting any more of my money. "Safeguarding our interests" I don't think! Paying your creditors is more like it.'

'Now look here, Freda, I've had enough of this.'

'So have I. But I'm not walking home so there's no point in stopping.'

'Then try getting this straight for a change—'

'Eric, there's that girl again.'

'What are you talking about? Anyone would think you'd got DTs.'

'Look—she's bending down to speak to you. She's trying to open your door.'

'Christ!'

'Eric, don't start the car like that. Don't drive so furiously. What are you trying to do?'

'I'm trying to outdistance her.'

'But the speed limit . . .'

'Damn the speed limit! What's the good of having a powerful car if you don't use it? . . . That's right. You hit the bottle again.'

'But the way you're driving! You ignored a halt sign. That lorry driver had to cram on his brakes.'

'What the hell! Look round and see if you can see her.'

'She's right behind us, Eric.'

'What, in her car?'

'No, she seems to be floating a little way above the ground. But she's moving fast. I can see her hair streaming out behind her.'

'Well, we're doing seventy-five ourselves.'

'But we can't go on like this for ever. Sooner or later we've got to get out.'

'Sooner or later she's got to get tired of this caper.'

'Where are we? This isn't the way home.'

'Do you want her following us home? I want to lose her. What do you take me for?'

'A bastard who's ruined my life and ended that poor girl's.'

'No one warned me you'd ruin mine. I wish they had. I might have listened. Warnings are only given to the deaf . . . Look again to see if Gisela's still following.'

'She's just behind us. Oh Eric, her eyes are wide and staring. She looks horribly, horribly dead. Do you suppose she'll ever stop following us? Gisela. It's a form of Giselle. Perhaps she's like the girl in the ballet and condemned to drive motorists to death instead of dancers.'

'Your cultural pretensions are impressive. Is your geography as good?'

'What do you mean?'

'I mean where the hell are we? I swear I've never seen this road before. It doesn't look like a road in southern England. More like the North Yorkshire moors, except that even there there's some habitation. Besides, we couldn't have driven that far.'

'There's a signpost at this next crossroads if you'll slow down enough for me to read . . .'

'Well?'

'I don't understand it, Eric. All four arms of the signpost are blank.'

'Vandals painted them out.'

'Vandals! In this desolate, isolated spot? Oh Eric, I don't like this. Suppose we're condemned to go on driving for ever?'

'No, Freda, the petrol would give out.'

'But the gauge has been at naught for ages. Hadn't you noticed?'

'What? So it is. But the car's going like a bird.'

'Couldn't you slow down a bit? I know you didn't for the signpost, but she—she's not so close behind us now . . . Please, Eric, my head's still aching.'

'What do you think I'm trying to do?'

'But we're doing eighty . . . I knew it. We'll have to go on driving till we die.'

'Don't be such an utter bloody fool. I admit we've seen a ghost—something I never believed existed. I admit I've lost control of this damn car and I don't know how she keeps running on no petrol. I also admit I don't know where we are. But for all this there's got to be a rational explanation. Some timeswitch in our minds. Some change of state.'

'That's it! Eric, what's the last landmark you can remember?'

'That blanked-out signpost.'

'Not that. I mean the last normal sign.'

'You said there was a halt sign, but I must say I never saw it.'

'You drove right through it, that's why. We shot straight in front of a lorry. I think—oh, Eric, I think we're dead.'

'Dead! You must be joking. Better have another drink.'

'I can't. The flask's empty. Besides, the dead don't drink. Or eat. They're like Gisela. You can't touch them. There's nothing there.'

'Where's Gisela now?'

'A long, long way behind us. After all, she's had her revenge.'

'You're hysterical, Freda. You're raving.'

'What do you expect but weeping and wailing? We're in Hell.'

'The religious beliefs of childhood reasserting themselves.'

'Well, what do you think Hell is? Don't hurry, you've got eternity to answer in. But I know what *I* think it is. It's the two of us driving on alone. For ever. Just the two of us, Eric. For evermore.'

The Call

ROBERT WESTALL

I'm rota-secretary of our local Samaritans. My job's to see our office is staffed twenty-four hours a day, 365 days a year. It's a load of headaches, I can tell you. And the worst headache for any branch is overnight on Christmas Eve.

Christmas night's easy; plenty have had enough of family junketings by then; nice to go on duty and give your stomach a rest. And New Year's Eve's OK, because we have Methodists and other teetotal types. But Christmas Eve . . .

Except we had Harry Lancaster.

In a way, Harry *was* the branch. Founder-member in 1963. A marvellous director all through the sixties. Available on the phone, day or night. Always the same quiet, unflappable voice, asking the right questions, soothing over-excited volunteers.

But he paid the price.

When he took early retirement from his firm in '73, we were glad. We thought we'd see even more of him. But we didn't. He took a six-month break from Sams. When he came back, he didn't take up the reins again. He took a much lighter job, treasurer. He didn't look ill, but he looked *faded*. Too long as a Sam. director can do that to you. But we were awfully glad just to have him back. No one was gladder than Maureen, the new director. Everybody cried on Maureen's shoulder, and Maureen cried on Harry's when it got rough.

Harry was the kind of guy you wish could go on for ever. But every so often, over the years, we'd realized he wasn't going to. His hair went snow-white; he got thinner and thinner. Gave up the treasurer-ship. From doing a duty once a week, he dropped to once a month. But we still *had* him. His presence was everywhere in the branch. The new directors, leaders, he'd trained them all. They still asked themselves in a tight spot, 'What would Harry do?' And what he did do was as good as ever. But his birthday kept on coming round. People would say with horrified disbelief, 'Harry'll be *seventy-four* next year!'

And yet, most of the time, we still had in our minds the fifty-year-old Harry, full of life, brimming with new ideas. We couldn't do without that dark-haired ghost.

And the one thing he never gave up was overnight duty on Christmas Eve. Rain, hail or snow, he'd be there. Alone.

Now alone is wrong; the rules say the office must be double-staffed at all times. There are two emergency phones. How could even Harry cope with both at once?

But Christmas Eve is hell to cover. Everyone's got children or grandchildren, or is going away. And Harry had always done it alone. He said it was a quiet shift; hardly anybody ever rang. Harry's empty log-book was there to prove it; never more than a couple of long-term clients who only wanted to talk over old times and wish Harry Merry Christmas.

So I let it go on.

Until, two days before Christmas last year, Harry went down with flu. Bad. He tried dosing himself with all kinds of things; swore he was still coming. Was *desperate* to come. But Mrs Harry got in the doctor; and the doctor was adamant. Harry argued; tried getting out of bed and dressed to prove he was OK. Then he fell and cracked his head on the bedpost, and the doctor gave him a shot meant to put him right out. But Harry, raving by this time, kept trying to get up, saying he must go . . .

But I only heard about that later. As rota-secretary I had my own troubles, finding his replacement. The rule is that if the rota-bloke can't get a replacement, he does the duty himself. In our branch, anyway. But I was already doing the seven-to-ten shift that night, then driving north to my parents.

Eighteen fruitless phone-calls later, I got somebody. Meg and Geoff Charlesworth. Just married; no kids.

When they came in at ten to relieve me, they were happy. Maybe they'd had a couple of drinks in the course of the evening. They were laughing; but they were certainly fit to drive. It is wrong to accuse them, as some did, later, of having had too many. Meg gave me a Christmas kiss. She'd wound a bit of silver tinsel through her hair, as some girls do at Christmas. They'd brought long red candles to light, and mince-pies to heat up in our kitchen and eat at midnight. It was just happiness; and it *was* Christmas Eve.

Then my wife tooted our car-horn outside, and I passed out of the story. The rest is hearsay; from the log they kept, and the reports they wrote, that were still lying in the in-tray the following morning.

They heard the distant bells of the parish church, filtering through the falling snow, announcing midnight. Meg got the mince-pies out of the oven, and Geoff was just kissing her, mouth full of flaky pastry, when the emergency phone went.

Being young and keen, they both grabbed for it. Meg won. Geoff shook his fist at her silently, and dutifully logged the call. Midnight exactly, according to his new watch. He heard Meg say what she'd been carefully coached to say, like Samaritans the world over.

'Samaritans—can I help you?'

She said it just right. Warm, but not gushing. Interested, but not *too* interested. That first phrase is all-important. Say it wrong, the client rings off without speaking.

Meg frowned. She said the phrase again. Geoff crouched close in support, trying to catch what he could from Meg's ear-piece. He said afterwards the line was very bad. Crackly, very crackly. Nothing but crackles, coming and going.

Meg said her phrase the third time. She gestured to Geoff that she wanted a chair. He silently got one, pushed it in behind her knees. She began to wind her fingers into the coiled telephone-cord, like all Samaritans do when they're anxious.

Meg said into the phone, 'I'd like to help if I can.' It was good to vary the phrase, otherwise clients began to think you were a tape-recording. She added, 'My name's Meg. What can I call *you*?' You never ask for their *real* name, at that stage; always what you can call them. Often they start off by giving a false name . . .

A voice spoke through the crackle. A female voice.

'He's going to kill me. I know he's going to kill me. When he comes back.' Geoff, who caught it from a distance, said it wasn't the phrases that were so awful. It was the way they were said.

Cold; so cold. And certain. It left no doubt in your mind he *would* come back and kill her. It wasn't a wild voice you could hope to calm down. It wasn't a cunning hysterical voice, trying to upset you. It wasn't the voice of a hoaxer, that to the trained Samaritan ear always has that little wobble in it, that might break down into a giggle at any minute and yet, till it does, must be taken absolutely seriously. Geoff said it was a voice as cold, as real, as hopeless as a tombstone.

'Why do you think he's going to kill you?' Geoff said Meg's voice was shaking, but only a little. Still warm, still interested.

Silence. Crackle.

'Has he threatened you?'

When the voice came again, it wasn't an answer to her question. It was another chunk of lonely hell, being spat out automatically; as if the woman at the other end was really only talking to herself.

'He's gone to let a boat through the lock. When he comes back, he's going to kill me.'

Meg's voice tried to go up an octave; she caught it just in time.

'Has he *threatened* you? What is he going to do?'

'He's goin' to push me in the river, so it looks like an accident.'

'Can't you swim?'

'There's half an inch of ice on the water. Nobody could live a minute.'

'Can't you get away ... before he comes back?'

'Nobody lives within miles. And I'm lame.'

'Can't I ... you ... ring the police?'

Geoff heard a click, as the line went dead. The dialling tone resumed. Meg put the phone down wearily, and suddenly shivered, though the office was over-warm, from the roaring gas-fire.

'Christ, I'm so *cold*!'

Geoff brought her cardigan, and put it round her. 'Shall I ring the duty-director, or will you?'

'You. If you heard it all.'

Tom Brett came down the line, brisk and cheerful. 'I've not gone to bed yet. Been filling the little blighter's Christmas stocking ...'

Geoff gave him the details. Tom Brett was everything a good duty-director should be. Listened without interrupting; came back solid and reassuring as a house.

'Boats don't go through the locks this time of night. Haven't done for twenty years. The old alkali steamers used to, when the alkali-trade was still going strong. The locks are only manned nine till five nowadays. Pleasure-boats can wait till morning. As if anyone would be moving a pleasure-boat this weather . . .'

'Are you *sure*?' asked Geoff doubtfully.

'Quite sure. Tell you something else—the river's nowhere near freezing over. Runs past my back-fence. Been watching it all day, 'cos I bought the lad a fishing-rod for Christmas, and it's not much fun if he can't try it out. You've been *had*, old son. Some Christmas joker having you on. Goodnight!'

'Hoax call,' said Geoff heavily, putting the phone down. 'No boats going through locks. No ice on the river. Look!' He pulled back the curtain from the office window. 'It's still quite warm out—the snow's melting, not even lying.'

Meg looked at the black wet road, and shivered again. 'That was no hoax. Did you think that voice was a hoax?'

'We'll do what the boss-man says. Ours not to reason why . . .'

He was still waiting for the kettle to boil, when the emergency phone went again.

The same voice.

'But he *can't* just push you in the river and get away with it!' said Meg desperately.

'He can. I always take the dog for a walk last thing. And there's places where the bank is crumbling and the fence's rotting. And the fog's coming down. He'll break a bit of fence, then put the leash on the dog, and throw it in after me. Doesn't matter whether the dog drowns or is found wanderin'. Either'll suit *him*. Then he'll ring the police an' say I'm missin' . . .'

'But why should he *want* to? What've you *done*? To deserve it?'

'I'm gettin' old. I've got a bad leg. I'm not much use to him. He's got a new bit o' skirt down the village . . .'

'But can't we . . .'

'All you can do for me, love, is to keep me company till he comes. It's lonely . . . That's not much to ask, is it?'

'Where *are* you?'

Geoff heard the line go dead again. He thought Meg looked like a corpse herself. White as a sheet. Dull dead eyes, full of pain. Ugly, almost. How she would look as an old woman, if life was rough on her. He hovered, helpless, desperate, while the whistling kettle wailed from the warm Samaritan kitchen.

'Ring Tom again, for Christ's sake,' said Meg, savagely.

Tom's voice was a little less genial. He'd got into bed and turned the light off . . .

'Same joker, eh? Bloody persistent. But she's getting her facts wrong. No fog where I am. Any where you are?'

'No,' said Geoff, pulling back the curtain again, feeling a nitwit.

'There were no fog warnings on the late-night weather forecast. Not even for low-lying districts . . .'

'No.'

'Well, I'll try to get my head down again. But don't hesitate to ring if anything *serious* crops up. As for this other lady . . . if she comes on again, just try to humour her. Don't argue—just try to make a relationship.'

In other words, thought Geoff miserably, don't bother me with *her* again.

But he turned back to a Meg still frantic with worry. Who would not be convinced. Even after she'd rung the local British Telecom weather summary, and was told quite clearly the night would be clear all over the Eastern Region.

'I want to know where she *is*. I want to know where she's ringing from . . .'

To placate her, Geoff got out the large-scale Ordnance-Survey maps that some offices carry. It wasn't a great problem. The Ousam was a rarity; the only canalized river with locks for fifty miles around. And there were only eight sets of locks on it.

'These four,' said Geoff, 'are right in the middle of towns and villages. So it can't be *them*. And there's a whole row of Navigation cottages at Sutton's Lock, and I know they're occupied, so it can't be *there*. And this last one—Ousby Point—is right on the sea and it's all docks and stone quays—there's no river-bank to crumble. So it's either Yaxton Bridge, or Moresby Abbey locks . . .'

The emergency phone rang again. There is a myth among old Samaritans that you can tell the quality of the incoming call by the sound of the phone-bell. Sometimes it's lonely, sometimes cheerful, sometimes downright frantic. Nonsense, of course. A bell is a bell is a bell . . .

But this ringing sounded so cold, so dreary, so dead, that for a second they both hesitated and looked at each other with dread. Then Meg slowly picked the phone up; like a bather hesitating on the bank of a cold grey river.

It was the voice again.

'The boat's gone through. He's just closing the lock gates. He'll be here in a minute . . .'

'What kind of boat is it?' asked Meg, with a desperate attempt at self-defence.

The voice sounded put-out for a second, then said, 'Oh, the usual. One of the big steamers. The *Lowestoft*, I think. Aye, the lock-gates are closed. He's coming up the path. Stay with me, love. Stay with me . . .'

Geoff took one look at his wife's grey, frozen, horrified face, and snatched the phone from her hand. He might be a Samaritan; but he was a husband, too. He wasn't sitting and watching his wife being screwed by some vicious hoaxer.

'Now *look*!' he said. 'Whoever you are! We want to help. We'd like to help. But stop feeding us lies. I know the *Lowestoft*. I've been aboard her. They gave her to the Sea-scouts, for a headquarters. She hasn't got an engine any more. She's a hulk. She's never moved for years. Now let's cut the cackle . . .'

The line went dead.

'Oh, *Geoff*!' said Meg.

'Sorry. But the moment I called her bluff, she rang off. That *proves* she's a hoaxer. All those old steamers were broken up for scrap, except the *Lowestoft*. She's a *hoaxer*, I tell you!'

'Or an old lady who's living in the past. Some old lady who's muddled and lonely and frightened. And you shouted at her . . .'

He felt like a murderer. It showed in his face. And she made the most of it.

'Go out and find her, Geoff. Drive over and see if you can find her . . .'

'And leave you alone in the office? Tom'd have my guts for garters . . .'

'Harry Lancaster always did it alone. I'll lock the door. I'll be all right. Go on, Geoff. She's lonely. Terrified.'

He'd never been so torn in his life. Between being a husband and being a Samaritan. That's why a lot of branches won't let

The Call

husband and wife do duty together. We won't, now. We had a
meeting about it; afterwards.

'Go *on*, Geoff. If she does anything silly, I'll never forgive
myself. She might chuck herself in the river . . .'

They both knew. In our parts, the river or the drain is often
the favourite way; rather than the usual overdose. The river
seems to *call* to the locals, when life gets too much for them.

'Let's ring Tom again . . .'

She gave him a look that withered him and Tom together. In
the silence that followed, they realized they were cut off from
their duty-director, from *all* the directors, from *all* help. The
most fatal thing, for Samaritans. They were poised on the verge
of the ultimate sin; going it alone.

He made a despairing noise in his throat; reached for his coat
and the car-keys. 'I'll do Yaxton Bridge. But I'll not do Moresby
Abbey. It's a mile along the towpath in the dark. It'd take me an
hour . . .'

He didn't wait for her dissent. He heard her lock the office
door behind him. At least she'd be safe behind a locked door . . .

He never thought that telephones got past locked doors.

He made Yaxton Bridge in eight minutes flat, skidding and
correcting his skids on the treacherous road. Lucky there wasn't
much traffic about.

On his right, the River Ousam beckoned, flat, black, deep and
still. A slight steam hung over the water, because it was just a
little warmer than the air.

It was getting on for one, by the time he reached the lock. But
there was still a light in one of the pair of lock-keeper's cottages.
And he knew at a glance that this wasn't the place. No ice on the
river; no fog. He hovered, unwilling to disturb the occupants.
Maybe they were in bed, the light left on to discourage burglars.

But when he crept up the garden path, he heard the sound of
the TV, a laugh, coughing. He knocked.

An elderly man's voice called through the door, 'Who's there?'

'Samaritans. I'm trying to find somebody's house. I'll push
my card through your letter-box.'

He scrabbled frantically through his wallet in the dark. The
door was opened. He passed through to a snug sitting-room, a
roaring fire. The old man turned down the sound of the TV. The

wife said he looked perished, and the Samaritans did such good work, turning out at all hours, even at Christmas. Then she went to make a cup of tea.

He asked the old man about ice, and fog, and a lock-keeper who lived alone with a lame wife. The old man shook his head. 'Couple who live next door's got three young kids . . .'

'Wife's not lame, is she?'

'Nay—a fine-lookin' lass wi' two grand legs on her . . .'

His wife, returning with the tea-tray, gave him a *very* old-fashioned look. Then she said, 'I've sort of got a memory of a lock-keeper wi' a lame wife—this was years ago, mind. Something not nice . . . but your memory goes, when you get old.'

'We worked the lock at Ousby Point on the coast, all our married lives,' said the old man apologetically. 'They just let us retire here, 'cos the cottage was goin' empty . . .'

Geoff scalded his mouth, drinking their tea, he was so frantic to get back. He did the journey in seven minutes; he was getting used to the skidding, by that time.

He parked the car outside the Sam. office, expecting her to hear his return and look out. But she didn't.

He knocked; he shouted to her through the door. No answer. Frantically he groped for his own key in the dark, and burst in.

She was sitting at the emergency phone, her face greyer than ever. Her eyes were far away, staring at the blank wall. They didn't swivel to greet him. He bent close to the phone in her hand and heard the same voice, the same cold hopeless tone, going on and on. It was sort of . . . hypnotic. He had to tear himself away, and grab a message pad. On it he scrawled, 'WHAT'S HAPPENING? WHERE IS SHE?'

He shoved it under Meg's nose. She didn't respond in any way at all. She seemed frozen, just listening. He pushed her shoulder, half angry, half frantic. But she was wooden, like a statue. Almost as if she was in a trance. In a wave of husbandly terror, he snatched the phone from her.

It immediately went dead.

He put it down, and shook Meg. For a moment she recognized him and smiled, sleepily. Then her face went rigid with fear.

'Her husband was in the house. He was just about to open the door where she was . . .'

'Did you find out where she was?'

'Moresby Abbey lock. She told me in the end. I got her confidence. Then *you* came and ruined it . . .'

She said it as if he was suddenly her enemy. An enemy, a fool, a bully, a murderer. Like all men. Then she said, 'I must go to her . . .'

'And leave the office unattended? That's *mad.*' He took off his coat with the car-keys, and hung it on the office door. He came back and looked at her again. She still seemed a bit odd, trance-like. But she smiled at him and said, 'Make me a quick cup of tea. I must go to the loo, before she rings again.'

Glad they were friends again, he went and put the kettle on. Stood impatiently waiting for it to boil, tapping his fingers on the sink-unit, trying to work out what they should do. He heard Meg's step in the hallway. Heard the toilet flush.

Then he heard a car start up outside.

His car.

He rushed out into the hall. The front door was swinging, letting in the snow. Where his car had been, there were only tyre-marks.

He was terrified now. Not for the woman. For Meg.

He rang Tom Brett, more frightened than any client Tom Brett had ever heard.

He told Tom what he knew.

'Moresby Locks,' said Tom. 'A lame woman. A murdering husband. Oh, my God. I'll be with you in five.'

'The exchange are putting emergency calls through to Jimmy Henry,' said Tom, peering through the whirling wet flakes that were clogging his windscreen-wipers. 'Do you know what way Meg was getting to Moresby Locks?'

'The only way,' said Geoff. 'Park at Wylop Bridge and walk a mile up the towpath.'

'There's a short cut. Down through the woods by the Abbey, and over the lock-gates. Not a lot of people know about it. I think we'll take that one. I want to get there before she does . . .'

'What the hell do you think's going on?'

'I've got an *idea*. But if I told you, you'd think I was out of my tiny shiny. So I won't. All I want is your Meg safe and dry, back in the Sam. office. And nothing in the log that headquarters might see . . .'

He turned off the by-pass, into a narrow track where hawthorn bushes reached out thorny arms and scraped at the paintwork of the car. After a long while, he grunted with satisfaction, clapped on the brakes and said, 'Come on.'

They ran across the narrow wooden walkway that sat precariously on top of the lock-gates. The flakes of snow whirled at them, in the light of Tom's torch. Behind the gates, the water stacked up, black, smooth, slightly steaming because it was warmer than the air. In an evil way, it called to Geoff. So easy to slip in, let the icy arms embrace you, slip away . . .

Then they were over, on the towpath. They looked left, right, listened.

Footsteps, woman's footsteps, to the right. They ran that way.

Geoff saw Meg's walking back, in its white raincoat . . .

And beyond Meg, leading Meg, another back, another woman's back. The back of a woman who limped.

A woman with a dog. A little white dog . . .

For some reason, neither of them called out to Meg. Fear of disturbing a Samaritan relationship, perhaps. Fear of breaking up something that neither of them understood. After all, they could afford to be patient now. They had found Meg safe. They were closing up quietly on her, only ten yards away. No danger . . .

Then, in the light of Tom's torch, a break in the white-painted fence on the river side.

And the figure of the limping woman turned through the gap in the fence, and walked out over the still black waters of the river.

And like a sleepwalker, Meg turned to follow . . .

They caught her on the very brink. Each of them caught her violently by one arm, like policemen arresting a criminal. Tom cursed, as one of his feet slipped down the bank and into the water. But he held on to them, as they all swayed on the brink, and he only got one very wet foot.

'What the hell am I doing here?' asked Meg, as if waking from a dream. 'She was talking to me. I'd got her confidence . . .'

'Did she tell you her name?'

'Agnes Todd.'

'Well,' said Tom, 'here's where Agnes Todd used to live.'

There were only low walls of stone, in the shape of a house. With stretches of concrete and old broken tile in between. There had been a phone, because there was still a telegraph pole, with a broken junction-box from which two black wires flapped like flags in the wind.

'Twenty-one years ago, Reg Todd kept this lock. His lame wife Agnes lived with him. They didn't get on well—people passing the cottage heard them quarrelling. Christmas Eve, 1964, he reported her missing to the police. She'd gone out for a walk with the dog, and not come back. The police searched. There was a bad fog down that night. They found a hole in the railing, just about where we saw one; and a hole in the ice, just glazing over. They found the dog's body next day; but they didn't find her for a month, till the ice on the River Ousam finally broke up.

'The police tried to make a case of it. Reg Todd *had* been carrying on with a girl in the village. But there were no marks of violence. In the end, she could have fallen, she could've been pushed, or she could've jumped. So they let Reg Todd go; and he left the district.'

There was a long silence. Then Geoff said, 'So you think . . .?'

'I think nowt,' said Tom Brett, suddenly very stubborn and solid and Fenman. 'I think nowt, and that's all I *know*. Now let's get your missus home.'

Nearly a year passed. In the November, after a short illness, Harry Lancaster died peacefully in his sleep. He had an enormous funeral. The church was full. Present Samaritans, past Samaritans from all over the country, more old clients than you could count, and even two of the top brass from Slough.

But it was not till everybody was leaving the house that Tom Brett stopped Geoff and Meg by the gate. More solid and Fenman than ever.

'I had a long chat wi' Harry,' he said, 'after he knew he was goin'. He told me. About Agnes Todd. She had rung him up on Christmas Eve. Every Christmas Eve for twenty years . . .'

'Did he know she was a . . .?' Geoff still couldn't say it.

'Oh, aye. No flies on Harry. The second year—while he was

still director—he persuaded the GPO to get an engineer to trace the number. How he managed to get them to do it on Christmas Eve, God only knows. But he had a way with him, Harry, in his day.'

'And . . .'

'The GPO were baffled. It was the old number of the lock-cottage all right. But the lock-cottage was demolished a year after the . . . whatever it was. Nobody would live there, afterwards. All the GPO found was a broken junction-box and wires trailin'. Just like we saw that night.'

'So he talked to her all those years . . . knowing?'

'Aye, but he wouldn't let anybody else do Christmas Eve. She was lonely, but he knew she was dangerous. Lonely an' dangerous. She wanted company.'

Meg shuddered. 'How could he bear it?'

'He was a Samaritan . . .'

'Why didn't he tell anybody?'

'Who'd have believed him?'

There were half a dozen of us in the office this Christmas Eve. Tom Brett, Maureen, Meg and Geoff, and me. All waiting for . . .

It never came. Nobody called at all.

'Do you think?' asked Maureen, with an attempt at a smile, her hand to her throat in a nervous gesture, in the weak light of dawn.

'Aye,' said Tom Brett. 'I think we've heard the last of her. Mebbe Harry took her with him. Or came back for her. Harry was like that. The best Samaritan I ever knew.'

His voice went funny on the last two words, and there was a shine on those stolid Fenman eyes. He said, 'I'll be off then.' And was gone.

The
Little Yellow
Dog

MARY WILLIAMS

In my mind I called him my sand-man, because I always saw him at bed-time, from my window, when my Aunt Daphne had left me and gone downstairs. He was a small greyish yellow man, like the beach itself in the twilight, when the sea and sky became one . . . merging towards the dim uncertain lines of dunes tufted with beards of rush.

My aunt's sea-side house stood high on the dunes, with only a small cluster of chalets and cottages straggling behind it to the village of Wyck-on-Sea. So I had a clear view from my window and I always knew exactly when the old man would appear . . . immediately after the church clock had struck eight. The tower of the church poked up from the left on the sea side of the hamlet, and I knew, with the queer instinct of children, that he came from there.

I was just seven years old, and the old man was my secret, like the little yellow dog. The dog, though, was my day-time companion. We played hide and seek in the sand-hills, and I only mentioned him once when my aunt came to find me for tea, and said, 'Who were you calling to, Johnny?'

'The little yellow dog,' I said. 'Look, there he goes.' He was racing ahead, his rear end, with its fuzzy funny tail quickly disappearing round a hump of the dunes. But my aunt who was peering closely, said, 'There's *no* little yellow dog. You're making things up again. You mustn't. It's really silly.'

She wasn't pretending. She just didn't see him. That's why I didn't tell her about the sand-man, because I knew it would be the same with him.

The shore at Wyck was wide and lonely, stretching for half a mile to the sea, when the tide was low, leaving just a few pools behind by occasional rocks and humps of mud.

Except for the dunes, everywhere round Wyck-on-Sea was flat. The roads and gardens seemed filmed always by thin sand, where poppies and star-shaped yellow daisy flowers grew in wild abandon. Dykes cut through the countryside, making a patchwork of fields rich with ripening corn and oats. There were butterflies too; hundreds of tiny blue butterflies flying and drifting on the hot air which was tangy with the smell of brine, sea-weed, sweet-briar, and the bitter smell of the yellow daisy.

It was all so long ago; yet the atmosphere of that particular stretch of the east coast is as vivid to me now as it was then; and I can still see in my mind clearly, that dancing, laughing, raggedy-looking pup, and the more mysterious figure of the sandy-looking old man as he passed each evening along the beach with his head turned up from his rounded back, his thin longish hair and beard blowing in the wind like the rushes of the dunes. Although I could not see his expression I knew he was looking for something. Once when I could not sleep I got out of bed and went to the window. He was returning from where he had been; his figure greenish gold in the moonlight, only more bent, as though he was saddened by great disappointment.

He walked . . . almost drifted along, with head bowed towards the glimmering sand, and when he reached the bend where the path led to the church, the shadows closed in on him and he was there no more.

The next day was sunny again; and when I told my aunt I was going to the beach, she looked at me doubtfully for a moment, then remarked, 'All right, Johnny. Yes, it's a lovely day. Later perhaps I'll join you; so don't hide and pretend you're playing silly games with your make-believe little yellow dog.'

I didn't promise; I just nodded and was presently running through a valley of dunes with the blue butterflies all round me, the yellow daisies smelling, and the sand warm on my bare toes.

The tide was half-way out, and I wandered about picking up razor-shells and some of the tiny pink ones that had holes in them, which I was collecting to give to my sister when I went home. She was recovering from chicken-pox, and would like them, I knew, if I could find enough for her to make a necklace.

I hadn't been out long when the little dog came racing towards me over a breakwater from the direction where the old man walked each evening. There was a funny little building there . . . a ruin . . . just under the sand hills, that someone said had been used in the war.

The little dog was whitey-gold from sand, and when he jumped up at me, laughing, I could see the sand on his tongue, in his eyes, and on the shaggy brows falling over them. He never barked; and this, I thought, was why my aunt didn't believe in him. But then, ours was a secret relationship, and barking would have given the show away.

'Come on . . . ' I called, starting to run with him beside me . . . 'You hide and I'll find you . . . '

Generally he bounded off when I said that, but this time he didn't; just turned back in the direction he'd come from, paused, looked at me, then went on again. It happened several times until at last I followed, a bit grudgingly, because the dunes thinned that way, and weren't nearly so good for hide-and-seek, merging eventually into a part called The Warren, half sand, half earth and grass, and riddled with rabbit holes.

There weren't so many blue butterflies there, and quite suddenly as we approached the ruin, the sun went in, leaving the air cool with a thin wind shivering from the sea. The little dog hurried ahead, but I knew he wasn't playing hide and seek any more, although once, for a few seconds his shaggy form became lost in the grey light, and I felt suddenly sad with the queer kind of loss only a child could feel . . . as though all the magic had gone for good . . . all the magic I'd known of those summer haunted hours with the little yellow dog.

Then I saw him again; a shadowed shape slipping into the darkness of the derelict tumbled doorway.

I went in after him. He looked round once then started digging with his two front paws; digging with a hungry urgency that I knew in some way must be terribly important.

'All right,' I said, thinking of buried treasure. 'I'll help.' And my brief depression seemed to lift a little.

How long we scraped in the sand and earth I don't know. There'd been an exceptionally high tide that night, which had sucked a good deal of ground away, leaving rubble exposed that could have been hidden for years. Great chunks of coastline

were being taken by the sea from time to time: My aunt had told me of a church and two empty cottages further on that had fallen and disappeared; that's why she didn't like me playing on the beach when the water was up.

I remembered this in my feverish attempts to help the little dog. He didn't seem to notice me any more, and he wasn't laughing, or playing, not even for a moment . . . just scrape, scrape . . . sniffing and scratching, until very gradually I began to get not only tired, but afraid, sensing intuitively the end of the adventure could be something less pleasant than pirates' gold.

So I got up, shook my clothes, and wiped the sandy dust from my eyes, knees and hands.

'I'm going,' I said. 'There's nothing there, anyway. It's a stupid game. I'm going home.'

I turned my head to look at him, but he wasn't there. He'd gone. I was sorry, and sad, and wanted to cry. He must have heard me when I was tidying and cleaning myself, and taken off without my noticing. I called and called, but there was no response. There was nothing left but the lonely shore outside the ruin, the far off sea which had turned from blue to grey, and the lonely trek back across the sands, which seemed bereft without the little yellow dog.

That afternoon early, I took my bucket and spade and when my aunt questioned me, told her I was going to dig on the beach. 'Don't be long then,' she said, 'the tide's turned. If you're not back by three I shall come and fetch you.'

I went out, and as I cut through the dunes the blue butterflies were there again, and the sun was warm, diamonded gold and silver under the brilliant sky. But there was no little yellow dog, and he didn't even come when I reached the ruined hut or whatever it was, and started to dig with my spade.

I was hot, and soon my shirt was sticking to my back with perspiration. I could feel rivulets of sweat trickling from my forehead over my eyes and down my face. But the place where I found it, at last, was cold and damp from rain and sea, and the thing was a shining white beneath the clinging rubble of dust.

I fetched some water in my bucket from a nearby pool, and threw it over the curled-up shape. Then I stood staring. I wasn't frightened . . . just awed . . . the skeleton in a queer kind of way was beautiful in its perfection of bone structure, lying there as if in a long sleep . . . the skeleton which could so easily have been that of a little yellow dog.

I moved it very gently a few inches to the door of the ruined building, went out, and then looked back. In the sunlight the bones glistened clear and pale, like ivory. Perhaps I cried a little then, I don't know. But after the brief pause I walked on towards my aunt's house, not turning, not wanting to see any more . . . grateful only for the sunshine and distant sound of the waves breaking, for being alive in a world of summertime filled with blue butterflies and starred clumps of yellow daisy flowers.

That night I watched from my window as usual, and saw the sandy old man walking along the shore. He was hurrying this time, with his head turned to the twilit sky . . . or perhaps it was the wind at his back that made me think so . . . the wind and the thin clouds of fine sand blown upwards towards the dunes.

I went back to bed; but I was restless and wakeful; and in about an hour . . . it *must* have been an hour . . . because I heard the church clock chiming nine . . . I got up and crossed to the window again.

The moon was just spreading its path of silver across the sea, gathering in radiance until the whole scene was a brilliant pattern of light and slipping shadows. It was then that I saw the old man returning, with something under one arm. Something that looked like a sack. And as he passed, the colour of the evening seemed to lift and change, momentarily bathing the bent figure in a quivering glow of rose.

Everything suddenly was mysteriously warm and comforting; and I knew then; knew something that was beyond understanding, or the need to understand; sensed also, that in a few years I would be beyond such knowledge, and must retain for as long as possible its strange mystical awareness.

A minute later the enhanced translucence of the sky faded once more into the pallor of moon-washed dusk. The figure of the old man with his burden slipped into the looming shadow of the church and everything was still and motionless, and curiously bereft.

Presently I went back to bed. I was tired, and slept well. When I awoke, the morning sun was already streaking through the curtains.

I got up, dressed and went downstairs, very quietly so that my aunt, who was in the kitchen, would not hear me. Then I let myself out of the front door, and made my way by the dunes, to the church. The gate was half open; I went through and up the path where the grave-stones stood on either side, emerging grey and chillingly remote from the grass, only tipped yet, with a glimmer of morning light; but the little yellow daisy flowers were there, and a few poppies shedding their scarlet petals on the faint drift of wind.

It did not take me long to find the old man's resting place. I recognized it from the curled up carefully arranged skeleton of the little yellow dog which lay innocently close to a mound of grass topped by a simple stone.

With a lump in my throat I went closer and read the epitaph.

Sacred to the memory of
William Thomas
born 1869 died 1939

And lower down, smaller, and more frailly inscribed:

In life he dearly loved his dog,
and died mourning him

Just then two tiny butterflies flew down from the sky. I held out my hand, and one fluttered and rested there for a second, velvet-bright in the morning dew.

Then I turned and went back to the house.

Later when I'd had breakfast, I returned to the churchyard. The pearly white skeleton no longer lay by the grassy mound. But a gardener was tidying up, and I wondered if he'd moved it away. It didn't matter because I felt everything was all right.

And that night I knew.

I saw them from my window, walking along the sands, just below the dunes. But the old man seemed taller, more erect, and gold in the golden light of evening, as gold as the little yellow dog trotting happily by his side. Once my sand-man stopped and threw a stick, and I watched the little dog bound on after it, laughing, I was sure, as he'd laughed with me. I stayed at the window staring after them until the fading sky enfolded them, leaving nothing behind but the wide expanse of beach below the sand-hills where the rushes blew.

I never saw them again; and no one else ever knew what had happened. In any case, no one really cared except me; the secret was mine alone . . . mine, the sand-man's and the little yellow dog's, and he had been lost in the dunes, rabbiting, probably.

Sometimes, even now, after so many years, I look back and remember, reclaimed by that other world of blue butterflies and yellow daisy flowers . . . the world of childhood, where dreams so often have a potency for transcending physical reality, and perhaps more of truth.

Who knows?

I, for one, am not prepared to answer.

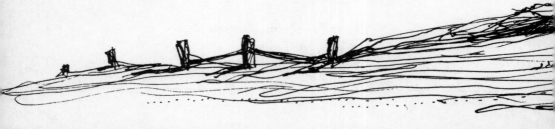

Acknowledgements

Vivien Alcock: 'The Rivals', ©1983 Vivien Alcock, first published in *Spooky: Stories of the Supernatural*, ed. Pamela Lonsdale (Thames/Methuen 1983), by permission of John Johnson Ltd. **Marc Alexander:** 'Spook House', copyright ©1985 Marc Alexander, from *Not After Nightfall* (first published by Viking Children's Books), by permission of Penguin Books Ltd. **Ruskin Bond:** ' The Monkeys', from *The Night Train at Deoli and Other Stories*, by kind permission of the author and Penguin Books India (P) Ltd. **Sydney J. Bounds:** 'Room at the Inn', ©1974 Sydney J. Bounds, published in *The 6th Armada Ghost Book*, ed. Mary Danby (Collins 1974). **Marjorie Bowen:** 'The Crown Derby Plate', from *The Last Bouquet* (Bodley Head 1933), by permission of Mr Hilary Long. **Redvers Brandling:** 'Mayday!', ©1994 Redvers Brandling, first published here by permission of the author. This story owes much to a real life incident when all the engines of a British Airways 747 failed in mid-air. The account is to be found in Betty Tootell: *All four engines have failed* (Deutsch). The cause was an intake of ash from an erupting volcano. In the crisis the crew performed magnificently, and no ghostly help was needed. **Petronella Breinburg:** 'It was Rose Hall', ©1975 Petronella Breinburg, from *The Haunted and the Haunters*, ed. Kathleen Lines (Bodley Head 1975), by permission of Random House UK Ltd. **Roberta Simpson Brown:** 'Earthbound' from *The Walking Trees and Other Scary Stories* (August House 1991). **Stephen Dunstone:** 'Fat Andy' ©1990 Stephen Dunstone from *The Man in Black: Macabre Stories from 'Fear on Four'* (BBC Books, 1990), by permission of Lemon, Unna & Durbridge on behalf of the author. **Jane Gardam:** 'Bang, Bang — Who's Dead?', copyright ©1987 Jane Gardam, from *Beware! Beware!*, ed. Jean Richardson (first published by Hamish Hamilton), by permission of Hamish Hamilton Ltd. **Adèle Geras:** 'Carlotta', ©1994 Adèle Geras, first published here by permission of the Laura Cecil Literary Agency. **John Gordon:** 'Little Black Pies' from *Catch Your Death* (Patrick Hardy Books 1984) by permission of A. P. Watt on behalf of the author. **Grace Hallworth:** 'The Guitarist', ©1984 Grace Hallworth, from *Mouth Open, Story Jump Out* (Methuen Children's Books 1984), by permission of Reed International Books. **Julia Hawkes-Moore:** 'The Chocolate Ghost', ©1994 Julia Hawkes-Moore, this version first published here by permission of the author. **Kenneth Ireland:** 'Children on the Bridge' from *We're Coming For You Jonathan*, © Kenneth Ireland 1983, (first published by Hodder & Stoughton, now in double volume Reed Consumer Books), by permission of the Jennifer Luithlen Agency on behalf of the author. **J. M. Johnson-Smith:** 'The Last Bus', first published in *Short Stories Magazine*, 1981. **Gerald Kersh:** 'The Scene of the Crime' from *Sad Road to the Sea* (Heinemann 1947). **Maria Leach:** ''Tain't So', from *Whistle in the Graveyard* (Viking Penguin 1974). Copyright © 1974 by Maria Leach, used by permission of Viking Penguin, a division of Penguin Books USA Inc. This story has been condensed by Maria Leach from a tale told by Emmy Seabrook in John Bennett: *Doctor to the Dead*, NY., Rinehart & Co. Inc. 1947. **Michael MacLiammoir:** 'The Servant' first published in the *Dublin Evening Herald*. **Jan Mark:** 'In Black and White', copyright ©1991 Jan Mark, from *In Black and White and Other Stories* (first published by Viking), by permission of Penguin Books Ltd. **Joyce Marsh:** 'The Warning', ©1978 Joyce Marsh, published in *The 10th Armada Ghost Book*, ed. Mary Danby

(Fontana 1978). **William F. Nolan**: 'Dead Call' from *Frights*, ed. Kirby McCauley, copyright © Kirby McCauley 1976, (Gollancz), by permission of Victor Gollancz. **Josh Pachter**: 'Caves in Cliffs', ©1984 Josh Pachter. **Philippa Pearce**: 'The Running-Companion', copyright ©1977 Philippa Pearce, from *The Shadow Cage and Other Tales of the Supernatural* (first published by Kestrel), by permission of Penguin Books Ltd. **Ann Pilling**: 'Gibson's', © Ann Pilling 1982, from *The 14th Armada Ghost Book*, ed. Mary Danby (Fontana 1982), by permission of Murray Pollinger on behalf of the author. **Alison Prince**: 'The Servant', from *The Green Ghost and Other Ghost Stories*, ed. Mary Danby (Collins 1989), by permission of HarperCollins Publishers Ltd. **J. J. Reneaux**: 'The Ring in the Rib' from *Cajun Folktales* (August House 1992). **Lennox Robinson**: 'A Pair of Muddy Shoes' from *Fear Fear Fear* (Ernest Benn). **Dal Stivens**: 'The Hard-Working Ghost' from *The Gambling Ghost and Other Tales* (Angus & Robertson 1954). **Barry Sutton**: 'The Tale of Caseley Halt' from *The Tenth Ghost Book*, ed. Aidan Chambers (Barrie & Jenkins 1974), by permission of Random House UK Ltd. **Rosemary Timperley**: 'Christmas Meeting', ©1952 Rosemary Timperley. **Michael Vestey**: 'An Apple for Miss Stevenson' from *The After Midnight Ghost Book*, ed. James Hale (Barrie & Jenkins 1980), by permission of Random House UK Ltd. **Elizabeth Walter**: 'Dual Control' from *Dead Woman and Other Haunting Experiences*, © Elizabeth Walter 1975, (Collins Harvill 1975), by permission of the author. **Robert Westall**: 'The Call', copyright ©1989 Robert Westall, from *The Call and Other Stories* (first published by Viking), by permission of Penguin Books Ltd. **Mary Williams**: 'The Little Yellow Dog', from *Chill Company: Ghost Stories from Cornwall* (Kimber 1976), by permission of Laurence Pollinger Ltd. on behalf of the author.
While every effort has been made to trace and contact copyright holders, this has not always been possible. If notified the publisher will be pleased to rectify any errors or omissions at the earliest opportunity.

The illustrations are by:
Nick Harris pp 190/191, 228/229, 236/237, 252, 256;
Jonathan Heap pp 14, 39, 41, 58, 90/91, 150, 265, 300/301;
Ian Miller pp 36, 98, 122/123, 131, 145, 173, 178, 196/197, 209, 216, 224;
Brian Pedley pp 7, 10/11, 46/47, 54, 136, 184, 240, 243;
Eric Stemp pp 81, 82, 116, 143, 166/167, 169, 261, 305, 310/311;
Adam Stower pp iii, 27, 31, 67, 72, 106/107, 159, 250, 269, 279, 285.
The cover photograph is by **Simon Marsden**.

'No,' said the man, 'they don't exist. I'd need some real, solid evidence before I could believe in ghosts.' He grinned. 'Real, *solid*, evidence.'

'I couldn't agree more.' His companion nodded vigorously. 'Now, if one were to vanish in front of your eyes, that would be real evidence.'

'Exactly,' said the man.

'Exactly,' his companion repeated, and vanished.